Energy Policy in the Emerging Economies

Energy Policy in the Emerging Economies

Climate Change Mitigation under the Constraints of Path Dependence

Sören Scholvin and Joachim Betz

LEXINGTON BOOKS
Lanham • Boulder • New York • London

Published by Lexington Books
An imprint of The Rowman & Littlefield Publishing Group, Inc.
4501 Forbes Boulevard, Suite 200, Lanham, Maryland 20706
www.rowman.com

6 Tinworth Street, London SE11 5AL, United Kingdom

British Library Cataloguing in Publication Information Available

Library of Congress Cataloging-in-Publication Data

Names: Scholvin, Sören, author. | Betz, Joachim, author.
Title: Energy policy in the emerging economies : climate change mitigation under the constraints of path dependence / Sören Scholvin and Joachim Betz.
Description: Lanham : Lexington Books, 2019. | Includes bibliographical references and index.
Identifiers: LCCN 2019003963 (print) | LCCN 2019005660 (ebook) | ISBN 9781498534376 (electronic) | ISBN 9781498534369 (cloth : alk. paper)
Subjects: LCSH: Energy policy—Developing countries. | Climate change mitigation—Developing countries.
Classification: LCC HD9502.D442 (ebook) | LCC HD9502.D442 B48 2019 (print) | DDC 333.7909172/4—dc23
LC record available at https://lccn.loc.gov/2019003963

Contents

List of Figures	vii
List of Tables	ix
Preface	xi
Acronyms	xiii
Introduction	1
1 China	27
2 India	47
3 Indonesia	69
4 South Korea	91
5 Brazil	107
6 Mexico	129
7 South Africa	145
Conclusion	167
References	193
Index	221
About the Authors	225

List of Figures

Figure 0.1	Primary Energy Consumption of the World's Major Energy Consumers, in quadrillion Btu	4
Figure 0.2	Energy Intensity of the World's Major Energy Consumers, in Btu per USD of GDP	5
Figure 0.3	Carbon Dioxide Emissions of the World's Major Energy Consumers, in million tonnes	5
Figure 0.4	Climate Change Vulnerability Index	10
Figure 1.1	China's Primary Energy Consumption by Source	28
Figure 2.1	India's Primary Energy Supply by Source	48
Figure 2.2	Energy-Related Infrastructure and Resources in India	51
Figure 2.3	Expansion of India's Electricity Generation Capacities, terawatt-hours (TWh) (business as usual)	55
Figure 2.4	Expansion of India's Electricity Generation Capacities, in TWh (ambitious)	55
Figure 3.1	Indonesia's Energy Generation by Source	70
Figure 3.2	Coal Resources and Power Stations in Indonesia	72
Figure 3.3	Oil and Gas Resources in Indonesia	74
Figure 3.4	Indonesia's Electricity Generation by Source	76
Figure 4.1	South Korea's Primary Energy Consumption by Source	92
Figure 4.2	South Korea's Present and Future Energy Demand by Source, in million tonnes of oil equivalent	97
Figure 5.1	Brazil's Primary Energy Generation by Source	108
Figure 5.2	Brazil's Electricity Generation by Source	109
Figure 5.3	Energy-Related Infrastructure and Resources in Brazil	112

Figure 6.1	Mexico's Primary Energy Consumption by Source	130
Figure 6.2	Mexico's Electricity Generation by Source	131
Figure 6.3	Mexico's Future Electricity Generation by Source	136
Figure 7.1	South Africa's Primary Energy Consumption by Source	147

List of Tables

Table 1.1 Chinese Overseas Oil and Gas Acquisitions from 2002 to 2013 41

Table C.1 Climate Change Mitigation by the Emerging Economies and Assessment Thereof 188

Preface

Almost on a weekly basis, both governmental and non-governmental organisations publish warnings on climate change and the failure of the international community to adequately mitigate it. The impact that rising temperatures and other alterations in the world's climate are expected to have in the coming decades should worry us indeed. Some negative effects have, in fact, already become troublesome reality.

Much has been published in order to explain why, despite all this, so little has been done to reduce the magnitude of climate change. Several researchers have developed roadmaps towards the climate and energy policies needed. This book makes a contribution to these debates, although the conclusions that we draw are not optimistic. In addition to the empirical facts presented here, the book advances an approach to studying climate and energy policies that relates to but is somewhat different from the academic mainstream. It is meant to complement energy transition studies and the concept of path dependence. We certainly recognise the insights gained from these more established perspectives, and are aware that adherents of energy transitions studies would in part concentrate on issues that are not central to our own analysis (while there is also considerable overlap). We hope that even readers with a different academic background and other convictions on how to do research on energy and climate policies will be open to our approach, and find this book to advance the state of the art.

When the manuscript was submitted to the publisher, its two authors had by then worked together for almost 12 years, jointly carrying out research on the emerging economies – especially on their climate and energy policies. Our first co-authored publication was a German Institute of Global and Area Studies (GIGA) Focus in 2007. GIGA Foci are semi-scientific papers of about 4,000 words length, also meant to serve as policy advice. In our own one, we compared the rapidly increasing energy demand of the emerging economies and explained resulting policies, with particular regard to the probability of conflicts over energy resources ensuing.

Single-authored papers on India and South Africa followed. Therein, we also concentrated on material conditions; this eventually turned into the approach that now marks this book. We first presented the framework for a comparative and structuralist analysis of the climate and energy policies of the emerging economies to an interdisciplinary research team directed by Daniel Flemes, at the GIGA. Our work also benefitted from discussions with PhD students such as Babette Never, who has pursued a quite different approach to the climate and energy policies of the emerging economies in her own research. We later organised a panel on these topics at the World Congress of Political Science in Montreal, Canada, in 2014, and also contributed to conferences and workshops in Germany and abroad, too. In 2016, we discussed draft versions of the introduction of this book and the chapter on South Africa with colleagues and students at Queen's University Belfast, Northern Ireland.

We are grateful for the numerous helpful comments received on these respective occasions. We would be particularly satisfied if the way that we now present the analytical framework and empirical analysis also managed to convince those who initially voiced scepticism about our research.

Sören Scholvin and Joachim Betz
Hamburg and Hanover, Germany

Acronyms

ANC	African National Congress
ANEEL	Agência Nacional de Energia Elétrica
APDRP	Accelerated Power Development and Reform Programme
APG	ASEAN Power Grid
ASEAN	Association of South East Asian Nations
ASGISA	Accelerated and Shared Growth Initiative
BATAN	Badan Tenaga Nuklir Nasional
Btu	British thermal unit
cbm	cubic metres
CCS	carbon capture and storage
CEPEL	Centro de Pesquisas de Energia Elétrica
CFE	Comisión Federal de Electricidad
CNOOC	China National Offshore Oil Corporation
CNPC	China National Petroleum Corporation
CTBE	Laboratório Nacional de Ciência e Tecnologia do Bioetanol
DEN	Dewan Energi Nasional
EIA	Energy Information Administration
Gasbol	Gasoduto Bolívia–Brasil
GEAR	Growth, Employment and Redistribution
GW	gigawatt
IEA	International Energy Agency
IIRSA	Iniciativa para a Integração da Infraestrutura Regional Sul-Americana
INDC	Intended Nationally Determined Contribution
IPCC	Intergovernmental Panel on Climate Change
IRENA	International Renewable Energy Agency
ITER	International Thermonuclear Experimental Reactor

KEPCO	Korea Electric Power Corporation
KETEP	Korea Institute of Energy Technology Evaluation and Planning
KOGAS	Korea Gas Corporation
kWh	kilowatt-hour
LNG	liquefied natural gas
MP3EI	Masterplan Percepatan dan Perluasan Pembangunan Ekonomi Indonesia
MW	megawatt
MWh	megawatt-hour
NDRC	National Development and Reform Commission
NEA	National Energy Administration
NEC	National Energy Commission
NELP	New Exploration Licensing Policy
NERSA	National Energy Regulator of South Africa
ONGC	Oil and Natural Gas Corporation
PBMR	Pebble Bed Modular Reactor
PDE	Plano Decenal de Expansão de Energia
PEMEX	Petróleos Mexicanos
Petrobras	Petróleo Brasileiro SA
PLN	Perusahaan Listrik Negara
PNE	Plano Nacional de Energia
PNMC	Política Nacional sobre Mudança do Clima
PPSA	Pré-Sal Petróleo SA
PROINFA	Programa de Incentivo às Fontes Alternativas de Energia Elétrica
PT	Partido dos Trabalhadores
REC	renewable energy certificate
REIPPPP	Renewable Energy Independent Power Producer Procurement Programme
RENE	Registro Nacional de Emisiones
RPS	renewable portfolio standard
RUPTL	Rencana Usaha Penyediaan Tenaga Listrik
SADC	Southern African Development Community
SAPP	Southern African Power Pool
SENER	Secretaría de Energía
SERC	State Electricity Regulatory Commission
SETC	State Economic and Trade Commission
Sinopec	China Petroleum and Chemical Corporation
TAGP	Trans-ASEAN Gas Pipeline Infrastructure Project
TWh	terawatt-hour

Introduction

This chapter begins with an overview of the coming shortage of fossil fuels and the enormous energy demand of the emerging economies. We show that these countries are critical to climate change mitigation because their carbon dioxide emissions are already large and will only continue to grow. This is mostly due to the ongoing and future expansion of generation capacities, which rests on coal, oil and gas.

At the same time, climate change will also hit the emerging economies hard. It already does so today, especially with regard to the decline of their agricultural output, the spread of diseases such as malaria as well as to flooding in coastal areas and along rivers. Based on an overview of the state of the art, we argue that there is need for comparative research that takes a rather global perspective on these matters. We summarise the approaches of energy transitions and path dependence, and show how they relate to and are complemented by our own presented analytical concept. This chapter closes with information on the diversity and independence of our cases, demonstrating how this book sheds light on causal mechanisms in different settings – which allows for a modest generalisation of the findings.

THE COMING SHORTAGE

Energy, the capacity to make things go, has always been essential to economics and politics. Jared Diamond (2005) explains in his book *Collapse* that the Norse, who had arrived on Greenland in AD 984, suffered from a serious lack of firewood after cutting down most of the island's trees. They were unable to generate enough energy to pasteurise their dairy products and melt iron ore to produce tools and weapons. Norse settlements on Greenland went extinct in the 1400s. Centuries later, Victorian Britain would build its global dominance on energy resources. The coal underneath the surface of England and Wales not only allowed the country to industrialise; it also enabled the Royal Navy to control the world seas with its steamships. The same logic applies to the United States in the twentieth century, and the oil wells now in the hands of US companies.

Today, the relevance of energy resources has all but diminished. Pointing at the rise of energy-thirsty powers from the Global South – in particular China and India – and the coming shortage of resources, Klare argues that "the pursuit and control of energy resources [has become] the central dynamic of world affairs" (2008: 7). In its *Energy Outlook 2035*, BP (2014) predicts that global energy consumption will rise by 41 per cent between 2012 and 2035. At the same time, depleting hydrocarbon resources will cause a decline in mining outputs. William Cummings, former spokesperson of ExxonMobil, and former Shell chairperson Ron Oxburgh argue that all easily extractable oil and gas resources have now been tapped. Many analysts, and the most important oil companies, expect that oil production will peak between 2010 and 2030, with a significant chance that the peak will be reached even before 2020 (Madureira 2014).

Because of the increasing difficulty to mine the remaining resources, Patzek and Croft (2010) suggested in 2010 that peak coal was due in the following year. Three years earlier, the Energy Watch Group (2007) – a network of politicians and scientists who are critical of fossil fuels and nuclear energy – calculated that peak coal would occur within 15 years. Mohr and Evans (2009) argue that the year 2026 is the likely one regarding the energy output of coal, while 2034 is the anticipated year concerning the quantity of coal. All these predictions are highly contested. According to Bentley (2002), the peak of natural gas production – with regard to conventional resources – is due in 2020. The Energy Information Administration (EIA 2014a) conversely expects natural gas production to increase up until the 2030s. Whereas Maggio and Cacciola (2012) project it to reach its apex in 2035, the Energy Watch Group (2013) reasons that peak production will be reached even before 2020. Regardless of the sometimes considerable variation regarding the dates predicted for the various resource peaks, the early twenty-first century is unquestionably marked by increasing difficulties to meet the demand for non-renewable energy.

At first glance, unconventional oil and gas seems to be a game changer – in particular for the United States. There, they have caused a re-industrialisation wave due to declining prices and turned the country into a net energy exporter (McKinsey 2017; Wang et al. 2014). However, experts warn that only 5 to 10 per cent of the shale gas found in the United States can be extracted at relatively low costs (Hughes 2013). These resources come from four major plays, which will reach their peak production in 2020 (Inman 2014). Unconventional oil, whose extraction is expected to peak in the late 2070s or early 2080s, cannot fully compensate for the coming shortage of conventional oil. Its existence merely slows down the decline of total oil production (Mohr and Evans 2010).

In other parts of the world, especially in Europe, civil society organisations that warn against the environmental side effects of the exploration and exploitation of unconventional oil and gas obstruct corresponding projects (*Frankfurter Rundschau*, 14 March 2014; *Guardian*, 8 March 2018). Even if environmentalists were appeased or alternatively excluded from the political process, it would still be impossible to extract a large part of the world's unconventional oil and gas at reasonable costs; also because these are marginal resources, meaning that their extraction will only be economically viable if the oil price is high and the sites of extraction benefit from a pre-existing infrastructure being in place – especially pipelines and storage facilities, a pool of qualified labour and arguably also tax incentives.

The coming shortage of energy resources will not be felt most strongly in the Global North. Energy intensity has been declining there due to increasing energy efficiency and saving, stagnant population growth and the shift from the industrial to the services sector. The Global North is, moreover, relatively well positioned to compensate for increasing energy prices. It has been relatively successful in switching to renewable energies. The countries that will be hit hardest by declining energy resources are, then, the emerging economies from the Global South, where the relevance of the industrial sector is only slightly declining (or, indeed, still on the rise) and where governments have subsidised energy consumption for decades – also to avoid social unrest.

India became the world's third-largest producer of electricity in 2013, with a share of 4.8 per cent in total global electricity generation (EIA 2016a). China's installed electricity generation capacity increased at an average annual rate of almost 12 per cent between 2000 and 2009. In 2013, the People's Republic had the largest thermal power capacity (945 gigawatt [GW]), the largest hydropower capacity (280 GW), the largest wind power capacity (91 GW) and one of the largest solar capacities (15 GW) in the world (EIA 2015d). Brazil, Indonesia, Iran, Mexico, Saudi Arabia, South Africa, South Korea, Thailand and Turkey are the other major energy consumers from the Global South, meanwhile.

Most of the emerging economies are more energy intense than the Global North (EIA 2015b), which results from the considerable relevance of the industrial sector for the former. Within this sector, energy-intense production is a particular strength: chemicals, pulp and paper, iron and steel, refining and non-metallic minerals. The energy intensity of the emerging economies is particularly problematic because fossil fuels – mostly coal – dominate their energy sectors. For example, 85 per cent of China's primary energy consumption is

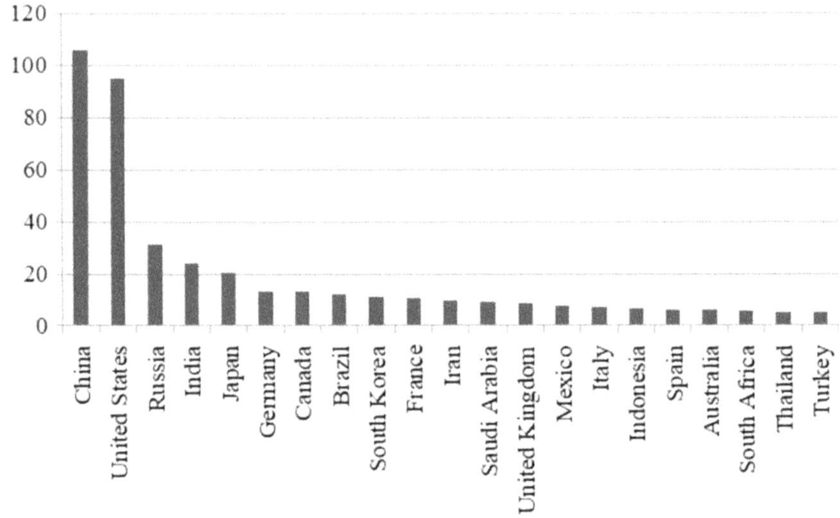

Figure 0.1 Primary Energy Consumption of the World's Major Energy Consumers, in quadrillion Btu. *Source*: EIA (2015a). *Note*: The British thermal unit (Btu) is a common measure for energy. It is defined as the amount of heat required to raise the temperature of one pound of water by 1°F.

based on coal, oil and gas.[1] The according value for Mexico is 93 per cent (EIA 2016b). Coal accounts for 94 per cent of the electricity generated by South Africa's state-owned power provider Eskom (EIA 2015g). This leads to significantly high carbon dioxide emissions. Considering the predicted strong growth of their respective gross domestic products (GDP) (PwC 2017), we argue that the emerging economies matter more for future energy consumption than the Global North does. This means that climate change mitigation depends to a large extent on the energy policies of these emerging economies.

Already today, the emerging economies face serious problems to meet their domestic demand for cheap energy. In Brazil, electricity prices increased considerably in 2012 and 2013 because of a reduction of subsidies and due to droughts that negatively affected the country's hydropower stations. Mexico would decide to raise state-controlled prices of electricity and natural gas on several occasions during the last couple of years. When the government increased petrol prices by 20 per cent in early 2017, countrywide protests incited looting that led to a number of deaths (*El Universal*, 1 and 6 January 2017). Indonesia spent more on fuel subsidies than it did on infrastructure and social welfare programmes combined between 2009 and 2013 (Beaton, Lontoh and Wai-Poi 2017). Massive subsidy cuts in the aftermath of the Asian Financial Crisis led to violent protests, which were one of the reasons for President Muhammad Suharto's eventual resignation from power.

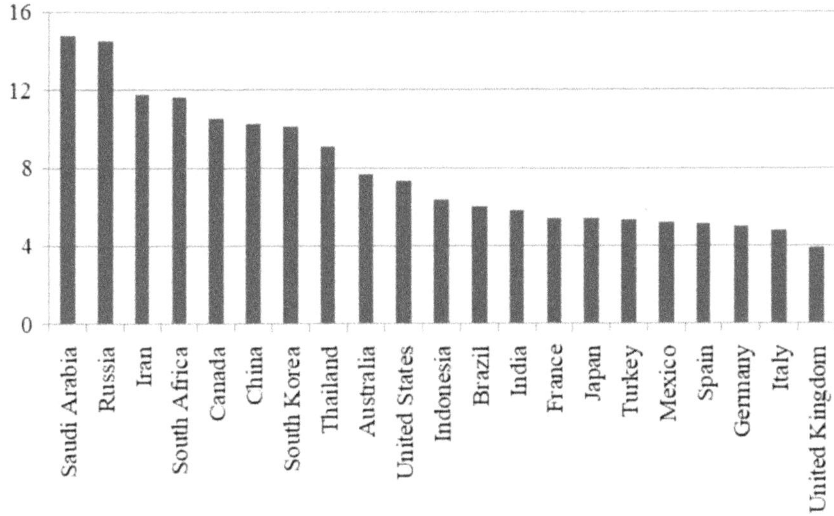

Figure 0.2 Energy Intensity of the World's Major Energy Consumers, in Btu per USD of GDP. *Source*: EIA (2015b).

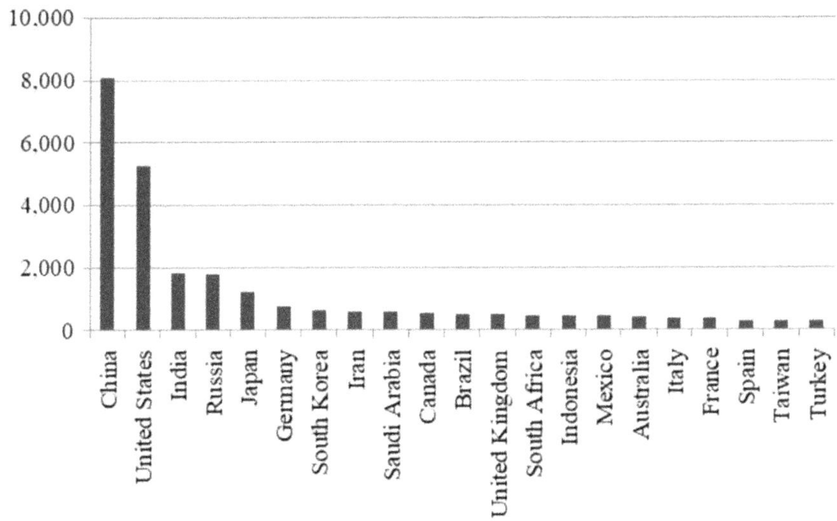

Figure 0.3 Carbon Dioxide Emissions of the World's Major Energy Consumers, in million tonnes. *Source*: EIA (2015c). *Note*: The values only represent carbon dioxide emissions that result from energy consumption.

India has recently abolished subsidies for diesel – because they contributed to an unsustainable fiscal deficit – and increased fuel taxes. In South Africa, load shedding became an everyday phenomenon in 2014 and 2015 (*Mail and Guardian*, 4 November 2014). South Koreans had to worry about such problems in 2013, when several nuclear power stations went offline (Reuters, 11 August 2013). In China, meanwhile, electricity shortage is not a major issue, but consumer prices are considerably high and the large fleet of coal-fired power stations account for dangerous levels of pollution in many urban areas even despite large-scale countermeasures being taken (*Economist*, 14 December 2017; GDP MAPS Working Group 2016).

EXPANSION STRATEGIES OF THE EMERGING ECONOMIES

Energy policy is, first of all, national self-help. In order to cover their growing energy needs, the emerging economies are expanding their domestic generation capacities. Brazil's National Ten-Year Energy Expansion Plan 2023 (Plano Decenal de Expansão de Energia 2023 [PDE 2023]), written before the country's severe recent economic and political crisis, projects an increase of oil production from 2 million barrels a day in 2013 to 4.9 million in 2023. Natural gas output will rise from 77.2 million cubic metres (cbm) a day to 148.8 million within the same time frame (Ministério de Minas e Energia 2014a). Electricity generation capacities are to rise from 125 to 196 GW. Mexico's Energy Ministry (Secretaría de Energía [SENER 2016]) expects electricity demand to rise by 3.4 per cent between 2016 and 2030. Power stations with a joint maximum output of 57 GW are to be built in the course of the next 15 years. At least 22 GW of these new capacities will be hydrocarbon-based, complementing a considerable renewables build-up dominated especially by wind power.

According to its Integrated Resource Plan (which is outdated but nevertheless still in force), South Africa will add 38 GW to its electricity generation capacities between 2010 and 2030, reaching a total of 85 GW by the latter year. By then, coal-fired power stations will still contribute the largest share: a total of 46 GW, or 48 per cent. Renewable energies are to expand considerably, though (Department of Energy 2011). South Korea's first national Energy Master Plan, which has been replaced by now and initially covered the years from 2008 to 2030, projected an increase of primary energy demand by nearly 50 per cent, equalling an annual growth rate of 3.7 per cent (Ministry of Trade, Industry and Energy 2014). The 8th Basic Plan for Electricity Supply and Demand, which presently guides the country's energy policy, foresees a somewhat lower increase in demand, however.

Indonesia plans to build new power stations capable of generating 80.5 GW by 2025 (Ministry of Energy and Mineral Resources 2016). The target energy mix for that year is 30 per cent coal, 25 per cent oil, 23 per cent hydropower and other renewables, and 22 per cent natural gas (President of the Republic 2014). In light of a rapidly increasing domestic electricity demand, the country embarked on a Fast-Track Electrification Plan in 2006. By now, 8.1 GW of coal-fired capacity have been added to the grid; another 1.9 GW are to follow. The second phase of nearly 18 GW of capacity additions include cleaner energy sources such as geothermal, hydroelectricity, natural gas and non-hydro renewables, although coal-fired power generation still accounts for the largest share.

According to somewhat extreme scenarios such as a report published by *Bloomberg News* (27 August 2013), China will need 88 GW of new capacities each year from now until 2030. The People's Republic will remain the world's largest coal consumer for the foreseeable future, accounting for half of global consumption today and an estimated share of slightly more than 40 per cent by 2040 meanwhile. The International Energy Agency (IEA 2016a) expects energy consumption in India to grow by 135 per cent between 2014 and 2040. Thermal power accounted for 58 per cent of planned capacity addition – a total of almost 79 GW – in the 11th Five-Year Plan. The more recent Draft National Energy Policy projects an increase of coal-based power generation capacity to 330 or even 441 GW by 2040. It suggests – albeit unrealistically – that, by then, renewables capacity will reach 597 to 710 GW (NITI Aayog 2017).

In sum, the energy policies of emerging powers are not characterised by a clear turn to renewables, which would be vital for meaningful climate change mitigation. Whereas some such as South Africa are pursuing quite ambitious goals with regard to renewable energies, others like India and Mexico concentrate on fossil fuels. As we show in this book, even the emerging economies that have announced impressive growth targets for renewable energies rather expand fossil fuel-based power generation in practice. Those rich in oil and gas – both conventional and unconventional – have developed (and partly already carried out) ambitious plans to exploit these resources.

International cooperation offers considerable prospects for many of the emerging economies. On the global scale, this is about nuclear power. China already possesses numerous nuclear power stations and its companies are experienced in nuclear technology. They build power stations not only at home but also abroad. The same applies to South Korea, whose government is seeking to turn the country into one of the main exporters of nuclear energy technology (World Nuclear Association 2017b). Mexico is the least advanced emerging economy in this regard, but its government is considering an expansion of the nuclear sector – adding power stations with a total capacity of up to 10 GW (SENER 2016). Mexico is also somewhat exceptional because it does not pursue any significant regional energy strategy, as Central America and the Caribbean offer few opportunities for cooperation on this matter.

South Korea's potential for regional cooperation on energy is practically non-existent for geographical reasons: its only land border is with North Korea and highly militarised. Efforts to acquire oil and gas resources abroad have not led to the envisaged results, meanwhile.

The other emerging economies have fostered international partnerships – particularly with Russia on nuclear energy – so as to gain access to key technologies. In 2014, Russia and South Africa signed an agreement on the procurement of nuclear power stations equipped with Russian reactor technology. South Africa has also signed cooperation agreements with China, France, South Korea and the United States. The Russian company Rosatom has shown interest in participating in the expansion of nuclear power in Brazil, as reported by the newspaper *O Globo* (7 June 2013). Given the lack of available public funds for building Brazilian-owned plants, Rosatom might itself build, own and operate nuclear power stations in Brazil, selling electricity at a guaranteed price. Rosatom also signed a memorandum of understanding with Indonesia's National Atomic Energy Agency (Badan Tenaga Nuklir Nasional [BATAN]) in 2015. In the previous year, an Indonesian–Russian consortium (with Rosatom participating therein) had been awarded a tender for the conceptual design of a 10 megawatt (MW) experimental reactor. Indonesia cooperates with Japan and South Korea, too. These partnerships have not, however, led to the construction of power plants. The first joint Indian–Russian project – the Kudankulam Nuclear Power Plant – has been generating electricity since 2013. The governments of the two countries announced in 2014 that at least 10 more nuclear reactors would be built by the Russians over the course of the next two decades (World Nuclear Association 2018).

At the regional level, some of the emerging economies have become drivers of bi- and multilateral cooperation on energy, although the ensuing projects look more ambitious on paper than what they eventually achieve. Brazil advances cooperation on hydropower and natural gas through the Initiative for the Integration of the Regional Infrastructure of South America (Iniciativa para a Integração da Infraestrutura Regional Sul-Americana [IIRSA]). The Bolivia–Brazil Pipeline (Gasoduto Bolívia–Brasil [Gasbol]) – which was first discussed in 1988 and finally turned operational in 1999 – connects Bolivia's natural gas resources to Brazil's south-eastern and southern regions. South Africa imports natural gas from Mozambique. Although unrealistic at the moment, the Kudu Gas Field in southern Namibia might be connected to consumers in the Western Cape, either by pipeline or by electricity transmission lines from a yet-to-be-built power station on the Namibian side of the border (News24, 25 November 2014). South Africa's state-owned electricity provider Eskom has agreed to import 2.5 GW from a planned third power station at the Inga Dam in Congo-Kinshasa (Maupin 2015). Indonesia is involved in the Trans-ASEAN Gas Pipeline Infrastructure Project (TAGP), which is to consist of 4,500 kilometres of pipelines mainly undersea, interlinking South East Asian countries. Eight bilateral pipeline projects with a total length of some 2,300 kilometres are already operational.

China is a key player in the scramble for the natural gas resources of Central Asia. Already in the late 1990s the considerable potential for regional cooperation on energy in that region was highlighted, with China being a potential major customer that could also further develop its own province of Xinjiang by building energy-related infrastructure in Central Asia (Dorian, Wigdortz and Gladney 1997; Dorian et al. 1999). Today, the Kazakhstan–China Pipeline runs from Xinjiang to the Caspian Sea. It connects to the Russian pipeline network, too. The Central Asian Gas Pipeline – which links China and Turkmenistan – was completed within just 28 months. What matters much more, though, is the bilateral efforts by Chinese state-owned oil companies to acquire resources worldwide for consumption in the People's Republic itself.

Indian enterprises have pursued the same objective, but with much less success. In 2012, the Indian government launched its Connect Central Asia Policy, signalling that it would foster economic and political relations with these states. Transregional energy infrastructure is a key component of this strategy (Dave 2016; Gupta 2013). India's state-owned Oil and Natural Gas Corporation (ONGC) has invested in Kazakhstan, but there appears to be no clear strategy on how to get Central Asian resources to India's domestic market – because of the Taliban insurgency in Afghanistan and political friction with Pakistan, India is cut off from Central Asia. A consortium of private companies has advanced plans to build an underwater pipeline for natural gas imports from Iran, bypassing Pakistan (*First Post*, 6 September 2017). Beyond hydrocarbons, India has constructed hydropower stations in Bhutan and is about to do the same in Nepal, too. These power stations deliver their output mostly to India, reaching a joint capacity in the lower single-digit GW range.

CLIMATE CHANGE AND ITS IMPACT

Very generally speaking, the largest impact of global warming is felt – and will continue to be so throughout this century – by countries close to the equator, ones that depend on agriculture, those suffering from water scarcities and by countries in danger of flooding (either along rivers or due to rising sea levels), too. These criteria apply particularly to Brazil, India and Indonesia, but the other countries in our sample also suffer disproportionally compared

Figure 0.4 Climate Change Vulnerability Index. *Source*: Authors' own draft.

to the rest of the world from climate change (World Bank 2013). The Climate Change Vulnerability Index indicates that large geographical areas within the emerging economies are marked by high or even extreme risks resulting from climate change, its impact on local people as well as their rather limited ability to adapt to it.[2]

The assessment based on the Climate Change Vulnerability Index is confirmed by what is already happening and what will continue to happen throughout this century in the emerging economies. In the Brazilian federal state of São Paulo, the frequency of heavy rainfall is expected to increase because of rising temperatures, especially in summer. The risk of floods and landslides, in particular in the rapidly growing shantytowns, will also rise. Higher temperatures will hamper the natural dispersion of pollutants in the city of São Paulo, which counts 11 million inhabitants and suffers from heavy traffic (Afonso Nobre et al. 2011). Crop yields are expected to decline significantly all over the country from now until 2050 as a consequence of global – and even more so of local – effects of climate change (Andersen et al. 2016). Climate change has already negatively affected agriculture in Mexico and will do so even more in the coming decades, with local temperatures rising by as much as 4°C and rainfall decreasing by up to 11 per cent (Galindo 2010). Every year, 1,035 square kilometres of farmland are lost as a result of climate change. More than 20 million Mexicans hence suffer from food insecurity (Deheza and Mora 2013). Some scholars suggest that up to 6.7 million adults may emigrate by the year 2080 because of declining agricultural productivity alone (Feng, Krueger and Oppenheimer 2010).

Pessimistic scenarios predict a decline of crop yields in South Africa by up to 90 per cent up until the end of this century. Some areas of the country have already become too hot for the cultivation of maize. Parts of the Free State and the Western Cape are no longer suitable for growing wheat (Benhin 2006). The former deputy director of South Africa's Department of Water Affairs, Mike Muller, warned in an article published in the *Mail and Guardian* (19 February 2011) that the country's metropolitan areas are heading for a water crisis by 2020 at the latest. Cape Town almost ran out of water in early 2018. The year 2015 was marked by El Niño – a natural phenomenon that causes disasters all across the Southern Hemisphere and whose frequency and intensity has been increased by climate change. As a consequence, the local supplier Rand Water suggested temporarily restricting water use in Gauteng (News24, 11 September 2015), South Africa's economic heartland. In KwaZulu-Natal, some 150,000 people would receive disaster aid and more than 40,000 head of cattle die (*Mail and Guardian*, 6 November 2015).

The *Fifth Assessment Report* of the Intergovernmental Panel on Climate Change (IPCC 2014) points out that the annual run-off of rivers in north

China will increase by up to 40 per cent in the coming decades. Shanghai – and other cities alike – will face a high risk of flooding. Rice yields are to diminish considerably in the east of the country. By the 2080s, maize yields in north China will probably have decreased by 25 per cent. Dengue fever may become endemic there. Béguin et al. (2011) predict that some regions in China will be exposed to a 50 per cent higher probability of malaria because of increasing humidity and temperatures.

Indonesia's capital Jakarta is one of the world's coastal cities most vulnerable to climate change (Yusuf and Francisco 2009). About 40 per cent of its surface is located below sea level. Floods in 2002 inundated nearly one-fifth of Jakarta's total area. A simulated projection by Susandi (2009) suggests that some parts of the city will be under water by 2050 if global warming continues at its current pace. In South Korea, a 25 per cent increase in reports of torrential rain and a 60 per cent increase in heavy rain warnings have been recorded during the last 20 years (Kim et al. 2002; Seo and Lee 2011). Much of the Korea Peninsula is expected to become a subtropical climate in the course of this century (Jeong 2010).

In South Asia, the monsoon will become heavier – being concentrated into fewer months each year. The Indian city of Surat, located in Gujarat and home to 4.5 million people, will soon be exposed to recurrent flooding, with the risk of epidemics of dengue and malaria ensuing in the aftermath (*Guardian*, 15 September 2014). Around seven million people will be displaced due to the submersion of parts of Chennai and Mumbai if global temperatures increase by a mere 2°C (Schroth, Helfer and Jude 2012). The *Times of India* (3 February 2007) reports that melting glaciers in Tibet will soon lead to severe floods in Assam. Climate-related factors may cause India's GDP to decline by as much as 9 per cent; also because the yields of major crops such as rice are expected to consequentially decrease by 40 per cent, too.

We are aware that the problems that metropoles in the Global South face are also due to poorly managed urbanisation, land subsidence as a result of groundwater depletion, the clearing of mangrove forests and to the straightening of rivers. South Africa's water crisis in 2015 has much to do with the fact that about half of the commodity is lost because of leaking pipelines (*Mail and Guardian*, 22 March 2012). In some places, indeed, climate change has positive effects. Yields in the Huang-Huai-Hai Plain – China's most productive wheat-growing region – are expected to increase, for example (IPCC 2014). Some of the newspaper articles and scientific publications to which we have referred in this section are admittedly alarmist, and perhaps overly pessimistic. Still, the balance of climate change is negative. Interacting with economic, political and social deficiencies, climate change poses a major threat to the prosperity and indeed stability of many societies around the globe – in particular, the emerging economies.

GLOBAL GOVERNANCE

Against this background, the international community has been working on binding agreements on the reduction of greenhouse gas emissions. A global governance framework for climate change mitigation is emerging and it has ambitious targets. At the United Nations Climate Summits in Copenhagen (2009) and in Paris (2015), numerous states, including the emerging economies, declared to what extent and specifically how they plan to reduce their greenhouse gas emissions. They did so in comparison to business-as-usual scenarios, meaning that even with their pledges it will still take decades for their emissions to decline in absolute terms (if they are to do so at all). The organisation Climate Action Tracker, whose assessments we summarise throughout this book, concludes that only India's commitment to climate change mitigation is currently sufficient to meet the 2°C objective. All other emerging economies contribute inadequately, with the respective commitments made by China, Indonesia, South Africa and South Korea being highly so – meaning that these four countries are among the worst performers worldwide.[3]

Energy policy is, of course, critical for climate change mitigation. Yet, contrary to other public goods that are also marked by strong external effects and in spite of the just-mentioned summits, there is no overarching global governance architecture for energy that can have a meaningful impact. National governments pursue strategies of national energy security, primarily by developing their own domestic resources and to a lesser extent by securing access to resources and technologies from abroad. Efforts to reduce energy intensity and increase the share of renewables are not framed by an effective global regime either. This situation calls for a serious effort to regulate energy policy, including competition over resources. The G20 appears to be the prime candidate for global energy governance, because it served as the main cooperation forum during the recent financial crisis. The members of the G20 did indeed agree at their summit in 2014 to work jointly on global energy governance, but it appears that major producers of energy resources are not keen to move ahead (Andrews-Speed and Shi 2016; Westphal and Röhrkasten 2016). Beyond the G20, there are several frameworks for global energy governance:

- The IEA, founded after the oil shocks of the 1970s, is the most influential body for international cooperation on energy. It works on emergency preparedness by providing strategic oil reserves and organises the exchange of information, for instance for the purpose of market analysis. Membership in the IEA is mostly confined to the Global North.
- The International Energy Forum was created as a dialogue platform between oil-consuming and oil-producing countries in 1991. It meets once a year, but has remained a data-gathering and deliberative body.

- The International Renewable Energy Agency (IRENA) was established in 2011, largely as a result of Germany's efforts to advance renewable energies. IRENA organises modest financial support for developing countries and also technology transfers, so as to boost renewables there.
- Beyond these large organisations, a range of new bodies for cooperation in specific areas of energy technology – for example on carbon capture and storage (CCS), energy efficiency and fuel cells and hydrogen – have been created. Their functions partly overlap, and there appears to be little coordination among them.

All in all, none of these bodies takes important political decisions on their own. They implement the will of their member states, insofar as they reach a consent. Against this backdrop, our analysis is limited to national policies on energy and climate change. We assess the emerging economies individually, shedding light on their international partnerships insofar as it matters to their respective energy and climate policies.

We are aware that there is an increasing amount of literature on the global governance of energy and even more so on climate change mitigation, with the books by Jordan et al. (2018) and Van de Graaf (2013a) being just two outstanding examples. Within this, particularly IRENA has been presented in a very positive light by scholars who argue that "the fascinating thing with IRENA is that new policy orientations and beliefs [. . .] can re-shape the landscape of energy" (Röhrkasten and Westphal 2013: 15). Van de Graaf (2013b) sees IRENA as a counterweight to the IEA's bias towards the fossil fuel and nuclear energy industries. Urpelainen and Van de Graaf (2015) point out, meanwhile, that the organisation stands out in terms of the timing of its creation, the speed of its ratification and the focus of its mandate. They suggest that IRENA can play a significant role in promoting the global diffusion of renewable energies. In other words, numerous scholars are convinced that global climate and energy governance is leading to the outcomes that everyone should sensibly wish for. From their perspective, such global governance is more than just the individual interests of states.

Yet, even some adherents of the liberal school of International Relations call such optimism into question. Most prominently, Keohane and Victor (2011) have introduced the term "regime complex for climate change". They conclude that there is little reason to expect that this complex will lead to reductions in emissions rapid enough to meet the 2°C goal. Indeed, it appears that even if the voluntary commitments made by most states in Paris in 2015 are kept, temperatures will rise by 2.6 to 3.1°C during this century (Rogelj et al. 2016).

Throughout this book, we show that the emerging economies have made international commitments to climate change mitigation that in most cases hardly limit their options for future energy-intensive growth. Even more worrisome, these commitments are usually contradicted by the ongoing expansion of fossil fuel-based energy generation – sometimes even by policy planning on this matter. Announcements made by Brazil's recently elected president, Jair Bolsonaro, a country to which we return in chapter 5, demonstrate that global governance does not hinder the emerging economies from taking back everything that they have previously promised on climate change mitigation. Such shifts in policy are not limited to erratic and radical politicians alone, as the example of Mexico's outgoing president Enrique Peña Nieto indicates. He has reversed the pro-climate approach of his predecessor and paved the ground for the large-scale extraction of conventional and unconventional hydrocarbon resources.

STATE OF RESEARCH

Besides government plans, which are usually available online, and the country briefs by the EIA, the availability of information and the state of research on energy policies and climate change differ considerably from one emerging economy to the next. Corresponding research on Brazil is focussed on the biofuels industry and its impact on greenhouse gas emissions and land use (Fargione et al. 2008; Macedo, Seabra and Silva 2008; Muylaert, Silva and Pires 2008; Pinguelli, Pires and Muylaert 2009). Opportunities for the expansion of solar and wind energy have been assessed (Bueno Pereira et al. 2006; Camargo do Amarante 2001; Martins and Bueno Pereira 2011; Ramos Martins 2008a, 2008b); so have climate change mitigation as well as governance and vulnerability related to the phenomenon (Afonso Nobre et al. 2011; Pereira Lindoso et al. 2011; Sarmiento Gutiérrez 2011; Seroa da Motta 2011; Viola and Franchini 2017). Particular attention has been paid to the role of the Amazon rainforest in this regard (Charles, Held and Nag 2013). Studies that deal with less prominent issues assess the expansion of nuclear energy and the prospects of tapping into Brazil's unconventional natural gas deposits (Carvalho and Sauer 2009; EIA 2013a; Kidd 2013; Patti 2015).

Prospects of renewable energy have also been studied for the case of Mexico (Gutiérrez Negrín 2012; Hernández Escobedo et al. 2014; IRENA 2015a, 2015b; Villicaña Ortiz et al. 2015). Energy efficiency remains a side issue in research on this country, although corresponding studies have generated surprising results (Rojas Cardenas et al. 2017). Privatisation and technological change in the oil sector is another topic of considerable interest. Since the 2013 Energy Reform, which partly liberalised the country's electricity as well as oil and gas sectors, numerous articles and books – including an *Energy Outlook* by the IEA (2016c) – have been published, describing the reform itself and also assessing its impact (Alpizar and Rodríguez 2016; Alvarez and Valencia 2016; Lajous 2014). The effect of climate change on migration and corresponding policy options for Mexico have been investigated, too (Deheza and Mora 2013; Feng, Krueger and Oppenheimer 2010).

Publications on South Korea concentrate on the shift towards green growth, including the promotion of renewable energies and the introduction of smart micro-grids (Choi 2015; Global Green Growth Institute 2015; In-Ha 2014; Jones and Yoo 2011; Kim and Mathews 2016; Mathews 2012). Park (2016) and Park et al. (2016) discuss the prospects for renewable energies, also as a substitute for nuclear energy – which plays a major role in South Korea's electricity mix. The overseas activities of South Korean energy companies have been studied, too (Jeon 2011).

Literature on China's energy policy is extremely broad, ranging from the global expansion of its oil companies and their influence on the national government to the impact of energy price shocks on the Chinese economy to the prospects of peak demand for coal occurring and its consequences (Buckley and Sanzillo 2015; Downs 2008; IEA 2014b; Müller-Kraenner 2008; Vermeer 2015; Wang and Zhu 2015; Wang and Zou 2014; Zhidong 2014). Alves (2015) shows how the engagement of the People's Republic with sub-Saharan African providers of energy resources differs from Brazilian activities in that particular region. Recent shifts in China's economic policy and their impact on energy consumption and greenhouse gas emissions have also been studied (Andrews-Speed et al. 2014; Green and Stern 2015, 2017; Hilton and Kerr 2017; Seligsohn 2011); so have the market prospects for renewables as well as the extraction of the country's shale gas resources (Gosens and Lu 2014; Gunningham 2014; Puttaswamy and Ali 2015). Research on climate change and China is equally exhaustive. It covers issues such as the approach of the People's Republic to international negotiations, cooperation with the United States, the potential of reducing emissions and energy intensity, geoengineering, as well as the influence of Chinese scientists and other actors on the government (Edney and Symons 2014; Kong 2011; Liu et al. 2015; Price, Wang and Yun 2010; Price et al. 2011; Wang 2015; Wang, Yang and Zhang 2015; Williams 2014; Wu 2013; Wübbeke 2013).

One of this book's authors has previously explained the role that Indian business associations played in the country's recent shift towards saving energy and promoting renewables (Betz 2012). Other scholars show, meanwhile, that the energy scenarios advanced by the Indian government are incompatible with emission targets (Singh 2011), mostly because the country faces the challenge of climate change mitigation under the condition of industrialisation on a much larger scale than the other emerging economies (Ebinger 2016). Against this backdrop, the numerous studies done on India's coal consumption are particularly relevant (Anand and Prasad 2013; Buckley and

Sharda 2016; Camilleri 2011). Research on renewables – for example the corresponding regulatory environment and the renewables' certificates market – is equally important and prominent (Goerild et al. 2014; Group Centennial 2009; Kaladharan 2016; Shrimali, Tirumalachetty and Nelson 2012). The *Handbook of Climate Change and India* (Dubash 2015) contains analyses of the impact of climate change, international negotiations and national politics. It also relates climate change to economic development. Further studies deal with the global activities of Indian energy companies and regional cooperation (Macdonald 2008; Müller-Kraenner 2008; Wijayatunga, Chattopadhyay and Fernando 2015). Drivers of change in India's and South Africa's climate governance have been compared by Never (2014). The consultancy firm Ernst & Young (2011, 2013) has published on the prospects of the oil and gas sector and of renewables in India.

With regard to Indonesia, opportunities for regional cooperation on natural gas and nuclear power have been scrutinised (Ahmed et al. 2017; Andrews-Speed 2016; Sovacool 2009); so has climate change mitigation (Brown and Peskett 2011; Hein 2013; Kaneko and Kawanishi 2016). The impact of energy subsidies on the Indonesian state's budget and related reforms are more prominent topics in the literature because of the impressive extent of these subsidies (Agustina et al. 2008; Beaton, Lontoh and Wai-Poi 2017; Braithwaite et al. 2012; Lontoh, Clarke and Beaton 2014; Lontoh, Beaton and Clarke 2015). The book *Decentralisation in Indonesia* (Turner et al. 2003) provides critical insights on institutional deficiencies that hamper an efficient energy policy; so does an article by Winters and Cawvey (2015) on geothermal energy. McKinsey (2014a) and PwC (2014) have published policy-oriented papers on Indonesia's energy sector in general and on the oil and gas sector in particular. Beyond that, the state of research – particularly on how energy policy and climate change interact – is thin to non-existent.

In South Africa, the Energy Research Centre at the University of Cape Town has published numerous policy-oriented studies, for instance on energy generation that would be compatible with mitigating climate change, the effectiveness of policies on renewables and lessons learnt from the country's renewable energy procurement programme (Eberhard, Kolker and Leigland 2014; Eberhard and Naude 2016; Edkins, Marquard and Winkler 2010; Winkler 2009; Winkler and Marquard 2009). This book's second author has published on the path-dependent development of South Africa's energy sector and the logic behind decision-making on regional cooperation (Scholvin 2014a, 2015). Cuesta Fernández (2015) and Maupin (2015) analyse the prospects of cooperation on electricity with Congo-Kinshasa and Ethiopia, respectively. The classical study by Fine and Rustomjee (1996) on the "minerals–energy complex" shows how business people and politicians involved in the energy and mining sectors shaped South Africa's economic policy before the watershed year of 1994. It is very unfortunate that there are presently no in-depth studies that apply Fine and Rustomjee's ideas to democratic South Africa, although there is politico-economic research on that country's decarbonisation (Baker 2015; Baker, Newell and Phillips 2014; Baker et al. 2015). The book *Blackout* (Styan 2015) provides a critical assessment of South Africa's energy policy, with particular regard to the state-owned provider Eskom.

What the previous literature overview indicates is that country-specific studies do not represent a major gap in the research, even though there are corresponding lacunae with regard to some of the emerging economies at least. However, most of the studies referred to above provide in-depth, country-specific analyses, ones usually not covering energy and/or climate policies in total but focussing on more narrow topics within these realms. What is most conspicuously missing is, rather, comparative research on the emerging economies that brings together energy policy and climate change in a conceptually guided, homogenous way. Edited volumes that address the effects of climate change on the emerging economies and/or their energy policies – for example *Climate Governance in the Developing World* (Held 2013) and *The Impact of Emerging Economies on Global Energy and the Environment* (Bagley, Moulioukova and Kassab 2015) – tend to consist of chapters that analyse these countries (and specific topics that are characteristic of them) one by one without truly engaging in comparison.

Explicitly comparative research that does exist addresses topics related but not identical to our own research interests. These scrutinised topics are ones rather narrow in scope: the liberalisation of the energy sector; increasing income as an opportunity for renewables; the emission-reduction potential of energy-efficiency measures; and, technology foresight (Proskuryakova 2017; Sadorsky 2009). Since the UN Climate Summit in Paris, climate change mitigation efforts by various countries have been ranked – but usually without scholars engaging in explanations of the policies behind these efforts (Aldy, Pizer and Akimoto 2017; Climate Transparency 2016; Compston and Bailey 2016). Somewhat similar comparative studies calculate global energy subsidies, meanwhile (Coady et al. 2015). An article by Never and Betz (2014) is one of the few exceptions, as research that is truly comparative and which succeeds in providing explanations instead of mere rankings.

As we outline in the last section of this introductory chapter, the comparative approach that we develop and apply here is suitable for verifying/falsifying key causal mechanisms. It enables us to explain, among other things, how the emerging economies will contribute to climate change mitigation. Before coming to our own approach, however, we first address energy transitions and path dependence – as somewhat related and predominant theoretical concepts.

ENERGY TRANSITIONS AND PATH DEPENDENCE

A frequently used approach to studying energy policy or energy transitions – the socio-technical one – deals with the interaction of three levels: namely, landscapes, socio-technical regimes and technological niches. Landscapes are the cultural, political and social contexts that mark societies at large (Geels and Schot 2007). Socio-technical regimes can be understood as prevailing sets of routines that create and reinforce a particular system embedded in institutions and infrastructures (Geels 2002). In other words, regimes reflect a predominant cognitive and physical mainstream of how something – including energy generation and transmission – is done. Technological niches allow for learning processes to occur, and for networks that support these innovations to be built up (Geels and Schot 2007; Kemp, Schot and Hoogma 1998). They are the starting point for major changes within socio-technical regimes.

Whether innovations spread beyond their niches and replace a regime depends on windows of opportunity that can only be explained by the regime itself – and, more importantly, the landscape. Windows of opportunity result from external pressure that destabilises a regime. Climate change constitutes such an external pressure. It works against the hydrocarbon-based regime and allows for renewable energies to emerge, potentially becoming a new regime. However, adaptation within the existing regime – such as the introduction of CCS technologies – is an alternative that is often easier than switching to an entirely new one (Smith, Stirling and Berkhout 2005). Regime persistence is boosted because energy ones are deeply interlinked with other socio-technical systems (Foxon 2011). For example, coal-based electricity generation has far-reaching implications for transport infrastructure. The resulting socio-technical transport system works against an energy transition, in particular one that rests on decentralised power generation. This means that interrelated socio-technical systems mutually reinforce each other (Rotmans, Kemp and van Asselt 2001), thus accounting for path dependence.

Focussing on the persistence of regimes reveals that mere technological solutions – for instance, the capacity to generate solar and wind energy at competitive prices – are not sufficient to mitigate climate change (Fouquet 2010; Geels 2014; Kuzemko et al. 2016). In the last decade, research on energy transitions has increasingly recognised that political power is absolutely critical. Some of the adherents of the approach now make use of literature from institutionalism and "varieties of capitalism" to explain the persistence of regimes or their replacements (Foxon, Köhler and Oughton 2008; Geels 2014; Kern 2011; Kuzemko et al. 2016; Mitchell 2008).

We acknowledge the insights gained from energy transition studies, but also think that the perspective therein neglects important factors concerning the emerging economies. The nourishing of niche innovations is unnecessary for the emerging economies, because applying renewable energies on a large scale is technically feasible nowadays. The poor performance and non-competitiveness of niche technologies, which has been highlighted by adherents of energy transition studies themselves (Jacobsson and Lauber 2006; Kemp, Schot and Hoogma 1998), do not apply anymore. What matters to the emerging economies is, rather, the large-scale implementation of renewables or – using the terminology of Rotmans, Kemp and Van Asselt (2001) – the take-off, breakthrough and subsequent stabilisation of these as a regime. Yet in energy transition studies, the state has mostly been conceptualised with regard to its support of niche technologies – meaning the pre-development of energy regimes (Foxon et al. 2005; Kern 2011; Raven et al. 2015; Smith and Raven 2012; Van den Bergh et al. 2007). What is more, the time frame within the literature on energy transitions is usually long, spanning over several decades (Fouquet 2010; Geels 2002; Geels and Schot 2007; Kemp, Schot and Hoogma 1998). Much less time is available for the emerging economies to switch to renewables, however, calling into question whether we can deduce lessons for them from studies of past energy transitions in the Global North.

Quite similar to energy transition studies, research that concentrates on path dependence stresses that it is extremely difficult to fundamentally change a country's energy matrix; not only because there is a complex socio-technical regime in place that encourages the maintenance of the status quo, but also because the existing energy regime is embedded in institutions and infrastructures (Unruh 2000, 2002). Hence, the essence of path dependence is that decisions taken in the past – for example on building a centralised power system with corresponding infrastructure – limit the policy options available today: "What happened at an earlier point of time will affect the possible outcomes of a sequence of events occurring at a later point in time" (Sewell 1996: 262–63). This means that change will occur incrementally, if at all, and it will take decades to alter an existing regime (David 1986).

In other words, path dependence is about the dynamics of self-reinforcing processes – or positive feedbacks – that make reversals difficult, at least once these processes have reached a certain intensity. The reason for this is that the relative benefits of the current activity compared with once-possible options will increase over time, even if the chosen technology becomes less efficient at a later point in it. In particular, energy systems are characterised by massive front-loaded financing and rapidly decreasing unit costs. Energy companies therefore shy away from investing in alternative technologies (such as renewables) as soon as they have developed and implemented a specific technology (such as fossil fuels). Beyond that, the proponents of established energy sources – that is, the proponents of the existing socio-technical regime – tend to be politically influential because they promise energy security, large-scale employment and significant tax contributions. This leads to a secondary lock-in that is neither economic nor technical, but rather political, in nature (Aghion et al. 2014).

What the concept of path dependence suggests with regard to energy policy is that the deployment of renewables is cost-intensive because of lock-in effects on fossil fuels. Corresponding changes cannot be effected by markets alone, because of the initial disadvantage of making an energy transition compared to business as usual. Government support – for instance, by introducing a carbon tax and supporting research on renewable energies with public funds – is needed up to the point where energy efficiency and renewables show increasing returns to scale (Acemoglu et al. 2012, 2016). The challenges of such a fundamental change have been highlighted by research on carbon lock-ins. The edited volume *Decarbonization in the European Union* (Dupont and Oberthür 2015) demonstrates that even countries from the Global North face major obstacles when adapting their energy sectors so as to mitigate climate change, because of lock-ins that result from the building sector, infrastructure for generating and transmitting power, general transportation logistics and even from foreign and trade policies.

However, the concept of path dependence also highlights that there are key opportunities for fundamental change at certain points in time. These are so-called critical junctures – a term first coined by Collier and Collier (1991), who conceptualise these junctures as turning points that allow for contingent choices that set a specific trajectory of path-dependent consolidation. For example, South Africa – facing sanctions because of its apartheid policies – sought to achieve energy autarky from the 1950s onwards. At this first critical juncture, the government decided to exploit the country's abundant coal resources, even launching a coal-to-liquids programme. The decisions taken about 70 years ago account for a development path that South Africa – being heavily dependent on coal, but having now reached a second critical juncture – may soon leave, as many of its coal-fired power stations are about to be decommissioned. Mexico, meanwhile, appears to have itself recently passed such a critical juncture, deciding to replace coal with natural gas.

A COMPARATIVE, STRUCTURALIST APPROACH

Our approach to the climate and energy policies of the emerging economies is focussed on structural factors, merging together what energy transition scholars call "landscapes" and the physical aspects of socio-technical regimes. Our assessment starts with material conditions. For instance, South Korea lacks hydrocarbon resources. Its potential for hydropower and non-hydro renewables is very limited, as we explain in chapter 4. Against this background, it is hardly surprising that the East Asian country has an advanced nuclear sector that accounts for 15 per cent of its primary energy consumption and comprises 25 reactors that reach a maximum output of 23 GW (EIA 2017).[4] Adding another example, South Africa runs short of oil and gas. It possesses huge amounts of coal underneath its surface, however. It is not surprising that this commodity dominates the energy sector: leaving aside private power providers, 94 per cent of South Africa's electricity is generated by burning coal (EIA 2015g).

We are not saying that the availability of resources determines a country's energy matrix. However, resource endowment constitutes a critical condition that accounts for path-dependent developments and we are convinced that it is not possible to understand the energy policies of the emerging economies without taking into consideration inherent challenges and opportunities. Yet, material conditions comprise more than natural resources (or the lack thereof). While most of the emerging economies pursue models of economic development that are based on energy-intensive industrialisation, China and South Korea appear to be willing to shift to less energy-intensive paradigms – as we explain in chapters 1 and 4, respectively. In addition to these structural features of the economy, a country's energy-related infrastructure is critical for options in energy policy. Reflecting basic ideas from studies of energy transitions and path dependence, we stress that all emerging economies face considerable challenges to decarbonise their respective energy sectors. In spite of the above-mentioned critical juncture, two giant coal-fired power stations presently under construction in South Africa will still be producing electricity in the second half of the twenty-first century. In China, there were about 12,000 coal mines up until recently (EIA 2015d). Efforts have been undertaken to close the least efficient ones, but the People's Republic will not just stop extracting and burning coal from one day to the next.

Against this backdrop, we summarise the energy demand projected for the emerging economies and the expansion plans for their respective energy sectors. Based on high and often overly optimistic expectations on GDP growth, the emerging economies seek to expand their power-generation capacities tremendously, pursuing ambitious objectives for renewables but also building new fossil-based capacities. We furthermore shed light on energy-related know-how, so as to find out whether the emerging economies even possess the technological capability to decarbonise their energy matrices. As explained above with regard to nuclear power, the emerging economies have been quite successful in acquiring technologies via foreign investments and international partnerships. This pattern also applies to renewable energies. China, meanwhile, has become a technology leader in renewable energies because of public investments in research and its huge domestic market. The People's Republic now accounts for more than 70 per cent of global solar cell production. Eight of the world's 15-largest producers of wind energy installation are Chinese (Gosens and Lu 2014; Puttaswamy and Sahil Ali 2015).

Sufficiency in terms of energy-related know-how does not necessarily facilitate decarbonisation, though. Some emerging economies, in particular Mexico and South Africa, have begun implementing CCS technologies, reinforcing the carbon path that they are already on. Brazil and Mexico foster the exploitation of oil and gas resources. This somewhat confirms the hypothesis from energy transition studies about adaption occurring within existing socio-technical regimes (instead of shifting to alternative ones). Others, in particular Indonesia, forgo opportunities to carry out research on renewable energy technologies and hardly make use of their naturally endowed potentials.

Another factor that is of interest to us is politico-economic constellations. South Korea's industrial sector has managed to slow down the government's efforts to decrease energy intensity (Global Green Growth Institute 2015). In China, public authorities below the national level often deliberately overshoot centrally determined targets because producing more energy and extracting more coal yields local tax benefits. Inefficient and high-polluting production facilities tend to be allowed to continue operating, as closing them down would decrease locally paid taxes and generate economic problems that the provincial authorities would have to deal with (Kong 2011; Williams 2014). South Africa's now moribund nuclear programme reflects corruption and self-enrichment, not economic rationality, as we summarise in chapter 7. In chapter 3, we explain how the decentralisation of political power in Indonesia works against the expansion of geothermal power because local authorities do not benefit from such projects; even if they are willing to support geothermal energy, they still lack the necessary capacities and means to facilitate foreign investment. In India, some federal states have hampered grid access for private electricity producers in the course of trying to protect the income of their own distribution companies (Singh 2016).

After having assessed material conditions, the energy demand and expansion plans for the energy sector, politico-economic constellations as well as energy-related know-how, we then analyse the climate policies of the emerging economies – drawing conclusions on their contribution to climate change mitigation. The emerging economies have pledged to reduce the carbon intensity of their economies considerably: for example, China by 60 to 65 per cent until 2030 and India by 33 to 35 per cent within the same time frame, both compared to business-as-usual scenarios. Some but not all of the emerging economies have specified when their carbon dioxide emissions will peak – between 2020 and 2025 in the case of South Africa, for instance.

Yet, referring to our previous analyses of the energy policies of the emerging economies, we show that it is rather unlikely that these goals – which have become part of the international climate change mitigation regime via each country's intended nationally determined contribution (INDC) – will be met. Considering the energy policies currently in place, most of the emerging economies will not contribute what environmentalist organisations such as Climate Action Tracker call a "fair share" to climate change mitigation. Indeed, even if the INDC are fulfilled, global warming in the course of this century will still be more than 2°C.

As noted, our analytical dimensions – material conditions, energy demand and expansion plans, politico-economic constellations, energy-related know-how as well as climate policy – serve as the basis for comparative research. As Sartori (1994) points out, comparison makes sense whenever the objects that are to be compared have something in common but are not identical. We study the seven emerging economies – all marked by rapid economic growth (with the exception of recent crises, most notably in Brazil), energy-intensive industrialisation, a strong reliance on fossil fuels and all subject to very serious consequences of climate change, among the other features that they have in common – on the basis of one single analytical concept. Doing so enables us to examine causal mechanisms in different contexts, which are to a certain extent independent from one another. Comparative research, as we understand it, stands between qualitative single-case studies and purely quantitative research. It is, hence, suitable for moderate generalisation, because it increases certainty regarding causal mechanisms by comparing whether they hold true across different and independent cases.

A few examples, on which we elaborate further in the respective country-specific chapters, demonstrate how diverse the contexts of the emerging economies are: Mexico has recently taken major steps towards liberalising its energy sector – a direction in which China and India move very hesitantly, if at all. China has become a technology leader, and is the second-largest economy on the planet. In this regard it is very different from the other countries here, whose national economies are much smaller (with the exception of India's) and whose technological capacities are rather limited. Brazil, Indonesia and South Africa possess reasonable chances of cooperating with their respective neighbouring countries on energy, which is not an option for Mexico and South Korea meanwhile.

The independence of our cases is limited by the fact that the decline in coal, oil and gas reserves is a global phenomenon. The still rapidly increasing hydrocarbon demand in India and the massive consumption thereof in China can be expected to put upward pressure on energy prices worldwide, making the provision of cheap energy difficult for the other emerging economies. Climate change and international agreements on its mitigation are, of course, also global phenomena that reduce the independence of the cases that we study here. Nevertheless, there are numerous factors that are independent across all cases: most importantly the respective domestic resource endowment and energy-related infrastructure, but also domestic politico-economic constellations. Mutually beneficial cooperation on energy in South America and sub-Saharan Africa, on the one hand, and the conflict-laden struggle for energy resources in Central Asia between China and India, on the other, are not related to one another either.

We have excluded Iran, Saudi Arabia, Thailand and Turkey from our analysis. Being two of the world's largest oil and gas exporters, Iran and Saudi Arabia suffer from their own wasteful consumption of energy rather than from depleting resources per se (Alyousef and Varnham 2010; Samsam Bakhtiari and Shahbudaghlou 2000; Woertz 2013). Their economies are built on revenues from oil and gas exports, meaning that the corresponding dynamics are totally different from those experienced by the emerging economies during the last two decades. Thailand and Turkey are excluded because of their relatively small GDPs and only moderate energy intensity in comparison to South Africa, which has a similarly small economy to both. We admit that one can make a sound case for including Thailand and Turkey in a comparative analysis of the energy and climate policies of the emerging economies; arguably without gaining additional insights on causal mechanisms, though. Yet, limitations of time to compose this book and space within it did not allow us to add these two specific cases.

NOTES

1. This information is available online at: www.tps-data-portal.org/all-datasets.
2. The Climate Change Vulnerability Index has been developed by the company Maplecroft to identify climate-related risks over the next 30 years, down to a level of 22 square kilometres worldwide. It does so by evaluating three components: exposure to extreme climate-related events, the sensitivity of societies and their adaptive capacities.
3. Country-specific reports are available online at: http://climateactiontracker.org/countries.html.
4. Further information is available online at: www.iea.org/statistics.

Chapter 1

China

In this chapter, we show how fossil fuels dominate China's energy sector. The country has an enormous potential for renewables, but this is only partly fulfilled because corresponding sites are not connected to the national power grid while unfavourable political conditions are also a hindrance. The latter range from an inadequate pricing system to an inefficient institutional setup for energy policy to the influence of coal and oil companies on the national government. Many provincial governments also support hydrocarbons because these reflect their own interests, at least in the short run. Still, the People's Republic has taken steps towards fundamentally changing energy consumption and, related to this, its entire economic model. We critically assess corresponding objectives from public-planning documents, paradigms such as the "new normal" as well as issue-specific policies and related instruments. We also elaborate how China seeks to secure energy resources abroad, provide an overview of its world leadership in renewables technology and summarise efforts to develop CCS and nuclear fusion. The chapter concludes with an assessment of China's climate policy, in particular its INDC – which is hardly ambitious, and highly insufficient as a fair contribution to climate change mitigation.

MATERIAL CONDITIONS

Coal is the main source of China's energy supply (66 per cent), followed by oil (22 per cent), natural gas (7 per cent), hydropower (3 per cent), other renewables (1 per cent) and nuclear power (1 per cent).[1] With regard to electricity generation, the dominance of coal is equally pronounced: it reaches a share of 63 per cent, followed by hydropower in a distant second place

(22 per cent). The share of coal in energy consumption has declined slightly during the last decade, whereas natural gas and non-hydro renewables have gained in importance – albeit at a comparatively low level. In 2013, the People's Republic had the largest thermal power capacity (945 GW), the largest hydropower capacity (280 GW), the largest wind power capacity (91 GW) and one of the largest solar capacities (15 GW) in the world (EIA 2015d).

Coal production in China peaked in 2013, declined slightly for the first time in 2014 and then even more so in 2015 (Buckley and Sanzillo 2015). The People's Republic will nevertheless remain the world's largest coal consumer for the foreseeable future, accounting for half of global consumption today and an estimated share of slightly more than 40 per cent in 2040. China is also by far the world's largest producer of this commodity, accounting for almost half of all global production. Its own domestic coal resources amount to 13 per cent of world reserves. Part of the coal that the People's Republic produces has a high ash and sulphur content. It is unsuitable for many applications. Reserves of high-quality coking coal – used most prominently for steel production – are scarce. Moreover, with the increasing need to extend the depth of mining, it is becoming ever more difficult and costly to exploit domestic reserves (IEA 2016a).

Beyond that, the efficiency of China's coal mines is low: the recovery rate stands at 30 to 35 per cent, far below best practice. This is partly due to the size of coal mines – small ones tend to use outdated equipment because of a lack of investment – and the fragmentation of production sites, as there were around 12,000 coal mines operating in the People's Republic up until a

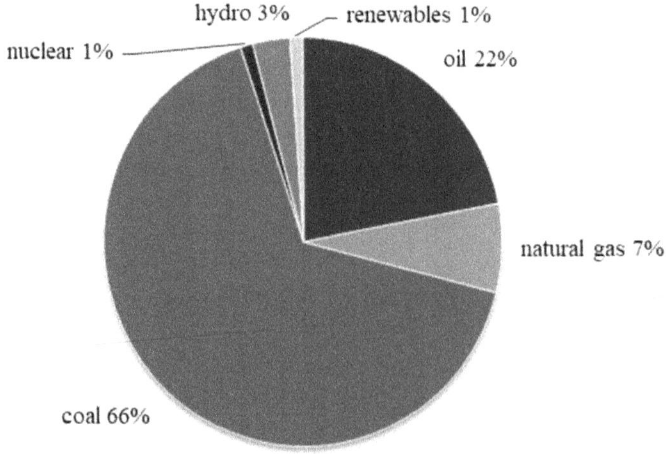

Figure 1.1 China's Primary Energy Consumption by Source. *Source*: Shift Project data, available online at: www.tsp-data-portal.org/all-datasets.

few years ago (EIA 2015d; IEA 2009). There is a considerable gap between these small operations on the one hand and the top state-owned companies on the other. The latter account for half of all production. Local state-owned companies produce another 20 per cent, while thousands of small mines contribute the remaining 30 per cent. As part of the 12th Five-Year Plan (2011–2015), the government started to consolidate the sector, to create larger mines and to close down hundreds of less efficient sites. In order to modernise large-scale mines and introduce new technologies, it opened up the coal sector to foreign investment in 2008. There was, initially at least, an enthusiastic response by foreign investors, which quickly faded away however because of convoluted and inconsistent regulations, and also because Chinese mining companies were (and still are) satisfied with the technology transfer that they had obtained, largely losing interest in joint ventures (Caprioni 2013).

Demand for coal is strongest in the coastal and southern regions, whereas supply remains concentrated in the northern and western provinces. Transporting it across several thousand kilometres has become a key challenge. It has made domestic coal uncompetitive for power production in coastal locations, boosting coal imports. It is therefore understandable that the Chinese government now concentrates on the development of its transmission network and the improvement of overstretched freight corridors. Building new generation capacities has been given secondary priority. Another important hindrance is water scarcity: most coal power plants are located in water-scarce regions. Water quotas have been applied to all provinces, making coal-fired power stations less competitive compared to other sources of electricity. Aside from these constraints, also significant is the fact that mining areas are highly vulnerable to environmental degradation on account of aridity. Being poorly developed in comparison to the coastal regions, China's north and west provinces thus lack the means to cope with the environmental impact of coal mining (Carbon Tracker Initiative 2014).

Having been a coal exporter for a long time, China has had to import the commodity since 2002 because of cost and quality advantages. The average cost of domestically mined coal has increased sharply since the year 2000, in spite of swift productivity gains. The main causes of this price hike are worsening geological conditions, new levies and taxes introduced to reduce the environmental impact of mining, transport bottlenecks, the massive restructuring of the industry and also the closing down of smaller mines (EIA 2015d; IEA 2010). Imports have climbed rapidly, making China the world's second-largest coal importer today. Australia and Indonesia are the major suppliers hereof, meaning that interruptions in the supply chain due to economic and political instability are unlikely to occur. Nonetheless, the Chinese government has launched schemes to switch to trading with partners from whom coal can be imported overland – most importantly, Mongolia.

The political dimensions of coal imports however remain marginal, unlike for imports of oil and gas.

As noted, oil is the second-largest source of energy. More than is the case with coal, oil demand is currently outstripping domestic production. China has proven reserves of 25 billion barrels, meaning that it ranks 14th globally. Eighty per cent of its production capacity is onshore in the north and north-east of the country. Active fields are ageing (EIA 2015d). New discoveries have not been made and production fell for the first time in 2016, as reported by the *Wall Street Journal* (25 August 2016). Offshore fields are small and rapidly maturing (EIA 2015d). Territorial disputes in the East China Sea (with Japan) and South China Sea (particularly with the Philippines and Vietnam) have hampered any large-scale investment. Hence, it is not surprising that the share of imports in China's oil consumption rose to 57 per cent in 2014. In that year, domestic production was 4.6 million barrels a day and consumption reached 10.7 million barrels, making the People's Republic the largest net oil importer worldwide. More than half of China's corresponding imports come from Middle Eastern countries. Africa also contributes a significant share of 30 per cent hereto. The two largest individual providers are Saudi Arabia (16 per cent) and Angola (13 per cent) respectively (EIA 2015d).

Although natural gas is not a major source of energy at present, its share in energy consumption is on the rise in accordance with government plans to substitute it for coal. China holds the largest proven natural gas reserves in the Asia-Pacific region, amounting to 4.6 trillion cbm, and was a net exporter up until 2007. Since then, imports have risen dramatically in tandem with rapidly proliferating domestic pipelines and processing infrastructure. Imports covered 32 per cent of demand in 2013 (EIA 2015d). The People's Republic holds the largest reservoir of recoverable shale gas worldwide: 31 trillion cbm. A major challenge with regard to both conventional and unconventional resources is that they are located far from the consumption centres, or even offshore. The coal-to-gas and shale gas industries are, further, at only an early stage of development because of technical challenges and scarce water resources as well as a lack of infrastructure. As a consequence, the ambitious targets for shale gas production – currently less than 1 per cent of total natural gas production – were halved in 2014. Bidding rounds open to private investors trailed behind schedule (Gunningham 2014; Green and Stern 2015). They were stopped altogether in 2016, meanwhile.

China has considerable potential for renewable energies, but one that is currently being fulfilled to only a limited extent. The technical potential of hydropower is 400 to 700 GW, of wind energy 1,500 to 2,800 GW and of solar energy 2,700 GW.[2] China has also abundant biomass resources, exploited only at 40 per cent capacity presently. The wind power potential is concentrated in the north-east (mostly offshore) and the north-west (onshore) of the country. Because of less cloud cover, solar irradiation is highest in the north and south-west. A key challenge to realising this tremendous potential for solar and wind power is overcoming the rather poor long-distance transmission system. As a consequence grid operators frequently curtail power generated from solar and wind installations, which run idle for significant periods of time (Cheung 2011; IRENA 2014).

Hydropower is already being exploited to a large extent at recently commissioned dams. China is by far the world's largest producer of hydroelectric power, with an installed capacity of almost 300 GW. The Three Gorges Dam, completed in 2012, is the largest hydropower station in the world. Another massive project, the Xiluodu Dam, entered into operation in 2013. Except for a few dams currently under construction, the remaining potential lies mainly in the underdeveloped wetlands of Sichuan, Tibet and Yunnan. These are suitable for small facilities (REN21 2015). Furthermore China has a large sector of biogas installations, with around 50 million units in operation, mostly used in farms – where they are integrated into livestock production. Liquid biofuels – largely based on ethanol – have not gained considerable relevance yet (IRENA 2014).

Concerning nuclear power, the aforementioned small share of it in overall energy consumption translates into an impressive total of 36 reactors that jointly generate up to 31.4 GW. A major challenge in this regard is that China's uranium resources that are recoverable at a competitive price of less than USD 260 a kilogramme are not overly large, totalling 272,000 tonnes – less than 4 per cent of overall global resources (NEA and IAEA 2016). This makes a shortfall in supply likely if the government's expansion plans do come true. The planned escape route is nuclear fusion (more on this later) and/or using thorium as a fuel, which is far more abundant in China's soil than uranium. However, the development of such technologies – also in Europe – has not proven economically viable yet. Due to its technical components, the enrichment process of thorium also comes along with a high risk of arms proliferation.

ENERGY DEMAND AND EXPANSION PLANS

China is a special case among the emerging economies; first of all because of the sheer volume of both its energy consumption and generation, which reach nearly twice the respective amounts of all our other cases combined. The People's Republic has contributed to about half of the worldwide growth in energy consumption and generation for a decade now. Its electricity generation capacity increased at an average annual rate of almost 12 per cent between 2000 and 2009. China's growth of energy consumption has slowed down by now, but this emerging economy is alone accountable for an impressive 23 per cent of global energy consumption and 26 per cent of global carbon dioxide emissions. By 2040, China will still be the largest energy consumer and its absolute carbon dioxide emissions will be twice as high as those of the United States (EIA 2015d; IEA 2017). Following somewhat extreme scenarios, the People's Republic will need 88 GW of new capacities each year from now until 2030 (*Bloomberg News*, 27 August 2013).

The second feature that sets China apart from the other countries analysed in this book is the fact that its industrial sector accounts for slightly more than 75 per cent of total energy consumption. China produces almost all types of manufactured goods – a consequence of many decades of self-sufficient and import-substituting industrialisation. Related to this, the People's Republic has become the "workbench" of the world, partly replacing low and medium capital- and skill-intensive production in the Global North. A considerable share of China's energy consumption and carbon dioxide emissions can be attributed to international trade (Hallding, Han and Olsson 2009; Liu et al. 2012). The preponderance of heavy industries used to be supported by concessional public credit, the provision of cheap land and by infrastructural services. Currency undervaluation has facilitated China's export drive, meanwhile.

Today, the movement into the "new normal" – as conceived by the government itself – not only means a successive reduction of the GDP growth rate but even more so a shift to less energy-intensive products and services (Meidan, Sen and Campbell 2015). The new normal also favours domestic consumption over exporting manufactured goods. This includes a focus on high-quality growth – meaning innovation, low-carbon technologies and the reduction of socio-economic inequality (Green and Stern 2017; Hilton and Kerr 2017). The new normal would allow China to honour its international commitments to mitigating climate change; or, in other words, what the National Development and Reform Commission (NDRC 2017) recognises as China's international responsibility. Since 2013, the share of industries in China's GDP – in particular, that of energy-intensive cement production, coal mining and steel manufacturing – has fallen (Green and Stern 2017).

The new normal marks a major turning point in China's economic and energy policy; however, public authorities have been attempting to bring energy consumption under control already since the early 1980s by the strict oversight of industrial energy use, financial incentives for efficiency-increasing investment and related services, as well as research and development (Price et al. 2011). Indeed, since 2002 the energy intensity of growth has declined from a very high level. In that year, a drastic shift occurred: senior leaders of the Communist Party and government agreed that future demand for energy had to be brought under control. A politburo communiqué called for a reduction in energy intensity by 20 per cent in the coming five years (Zhou et al. 2011). This target was then incorporated into the 11th Five-Year Plan (2006–2010).

An absolute increase in energy consumption by 4 per cent a year was planned for the time period covered by the National Climate Change Programme, together with a reduction of the share of coal in primary energy consumption by 3 per cent and of oil by 0.5 per cent. The shares of natural gas, hydropower and other renewables were projected to increase by 2.5, 0.6 and 0.3 per cent, respectively. The reduction of energy intensity was to be achieved by structural change, optimising industrial structures and, most importantly, by reducing the share of energy-intensive industries. Energy-conservation technologies were to be developed and disseminated. Specifically targeted sectors were iron and steel production, non-ferrous metal production, chemical industries, electricity generation and coal mining. The programme also aimed at enhancing the fuel efficiency of vehicles and phasing out energy-inefficient ones. It encouraged the marketing of energy-efficient products. The national energy-saving target was broken down to the provincial – and, at a later stage, local – level, allowing for a certain degree of variation among municipalities and provinces. Bureaucrats not achieving set targets were threatened with not being promoted (NDRC 2007).

The National Climate Change Programme was accompanied with several more specific programmes launched after 2005 in order to boost energy efficiency. The Top-1,000 Programme set energy-saving targets for China's largest individual energy consumers across nine industries. These companies were required to commission energy audits and to report on them quarterly. They then benefitted from various funding opportunities, the generosity of which was dependent on the amount of energy saved (Price, Wang and Yun 2010). This programme has by now been extended, covering the 10,000 largest energy consumers. The Ten Key Projects Programme, meanwhile, aims at retrofitting coal-fired industrial boilers, promoting combined cycle heat and power generation at the district level and also waste-to-heat utilisation. It supports measures to conserve oil, and to substitute it. Energy systems are to be optimised, as are vehicle engines and the energy efficiency of buildings. The Small Plant Closure Programme is devised as a way to optimise the structure of China's manufacturing sector, and therefore offers financial incentives to help ease the social costs of closing outdated factories (Price, Wang and Yun 2010; Price et al. 2011; Song et al. 2015).

The Renewable Energy Law, enacted in 2005, sets the stage for the scaling-up of renewable energies and obliges grid companies to purchase all electricity generated from renewable sources at favourable prices. The Medium- and Long-Term Development Plan for Renewable Energy from 2007 outlines China's commitment to increasing the share of renewables to

15 per cent of the primary energy supply by 2020. Under the Golden Sun Programme, announced in 2009, the government declared that it would subsidise the costs of installing solar power generation and transmission systems for projects selected by provincial governments by up to 70 per cent. Expansion plans for hydropower, meanwhile, suffered from a setback: targets for this type of power were revised down from 420 to 350 GW by 2020, per the 11th Five-Year Plan. The reason for this is that the best sites for hydropower have already been utilised.

It appears, however, that all of the measures mentioned so far were relatively easy to put into practice. At the beginning of this decade, the People's Republic reached a point at which energy savings became progressively more difficult. Hence, it comes as no surprise that the 12th Five-Year Plan did not demonstrate greater ambitions than its predecessor. For the duration of the 12th Five-Year Plan, the energy intensity of production was to drop by 16 per cent. The share of non-fossil fuels in the energy mix should increase to a total of 11.4 per cent, compared to 8.6 per cent in 2010. The projected growth in energy consumption was raised in comparison to the previous plan: namely, to 4.3 per cent a year. Coal production was projected to increase by 4.8 per cent a year. Consumption of natural gas and renewable energies were to grow by an annual 10.5 and 10.8 per cent respectively, indicating a diversification of energy sources.

The 12th Five-Year Plan moreover determined that the aforementioned Top-1,000 Programme be extended to cover a total of 10,000 firms. Rail transport was favoured over road transport, that by the construction of another 35,000 kilometres of high-speed railway lines and the expansion of urban transit systems. Further to this plan, a thorough shift in industrial policy was announced so as to promote the biomass, nuclear, solar and wind technology industries, as well as to support developers and manufacturers of hybrid and electric vehicles (Lewis 2011; Seligsohn 2011; Seligsohn and Hsu 2011).

Unlike many of the other emerging economies, China usually does actually meet its targets on energy policy. Energy intensity declined almost at the rate projected by the 11th Five-Year Plan, although only due to rationing at the end of the time period covered by it. Most of the specific programmes also met their aims (Price et al. 2011). Retrofits of existing buildings, the completion of metering reform and increasing the energy efficiency of household appliances fell, however, short of targets – partly because the incentives for private households were insufficient, as Price et al. (2011) argue. Building requirements do not apply to rural areas. Many targets set by the 12th Five-Year Plan were surpassed, as the IEA (2016b) and Ma (2015) summarise, and there is little doubt that China will indeed honour its commitments made at the UN Climate Conference in Paris (more on this later). The People's

Republic has also made significant progress in increasing the efficiency of thermal power generation, a consequence of shutting down small generators and replacing them with larger, supercritical and ultra-supercritical units.[3] Eighty-five per cent of all such plants built worldwide in 2014 are situated in China, meaning that almost half of its coal-fired power stations are by now technologically as advanced as – and therefore also as efficient as – possible (IEA 2015a).

Renewables, meanwhile, have been slowed down by curtailment, meaning that less renewable energy was purchased than produced. Curtailment rates for wind power ranged from between 7 and 17 per cent in the first half of this decade. They can be explained by the intermittent nature of renewable power generation and transmission difficulties, but also result from obstruction by local authorities – who prefer the continued expansion of fossil fuels. The reason for this is that local bureaucrats are evaluated on achieving economic growth, rather than environmental, objectives. Bureaucrats also consider large fossil fuel projects easier to implement than renewable ones. Moreover targets for renewables are defined in terms of capacity addition, not de facto generation. Thus, there is no incentive to initiate operations (Kostka 2014; Van Aken and Lewis 2015). Many local authorities have by now been directed by the central government to fulfil their renewable targets. They are hindered from continuing with the construction of new coal-fired power plants (Song et al. 2015). The planned reduction of the feed-in tariff for wind power and efforts to consolidate China's solar industry, presently plagued by overcapacity, constitute further challenges for the greater expansion of environmentally friendly electricity generation (REN21 2015).

With regard to today's energy policy, the Energy Development Strategic Action Plan (2014–2020) sets the road map. It calls for energy-related innovation, saving energy, strengthening domestic supply and increasing the share of renewables. The plan sets a target of total energy consumption of 3.36 billion tonnes of oil equivalent by 2020,[4] implying a modest annual growth rate of 2.2 per cent. A cap on coal consumption is set at 4.2 billion tonnes, meaning that annual production must decline by 2.9 per cent and the share of coal in the energy mix needs to drop to 62 per cent.

The 13th Five-Year Plan (2016–2020) reiterates the cap on coal. It sets a target of 15 per cent for the share of renewables, including hydropower, in the energy mix by 2020. This equals a total contribution by hydropower of 350 GW, by solar power of 100 GW and by wind power of 200 GW. Nuclear power should also expand rapidly: namely, to a total generation potential of 58 GW by 2020. Energy efficiency is to be boosted by cutting overcapacity, especially in coal mining and steel production. Quite impressively, the construction of new power plants was halted in 15 regions that were experiencing power oversupply in 2016 (Seligsohn 2016; Tianjie 2016). With regard to the more distant future, China plans to increase the share of non-fossil fuels in its energy mix to 20 per cent by 2030 and to 60 per cent by 2050.

POLITICO-ECONOMIC CONSTELLATIONS

The mapping out of China's energy policy and the clarity of its corresponding course suffered in the past from the fragmentation of state-owned enterprises, the relative policymaking autonomy of the provinces and energy companies, considerable administrative overlap and only rather modest success in overall coordination. There was no unified institutional energy set-up for a long time, in spite of the strict command exercised by the Communist Party. Although state-owned energy companies were formally brought under the jurisdiction of the newly formed Ministry of Energy in 1988, this government body failed to govern effectively. Its authority overlapped with that of the State Planning Commission, while the ministry's influence was confined only to the electricity sector.

As part of a fundamental reorganisation of government agencies, the State Economic and Trade Commission (SETC) was given oversight of the whole energy sector in 1998 – including of the state-owned energy companies. However, until today this agency has remained without either the manpower or indeed the clout to regulate the industry effectively. The vacuum has been filled instead by national oil companies as agenda setters, a turn of circumstance that works against major shifts in energy policy. These companies have been granted the right to trade freely at home and abroad and – in principle – to make their own investment decisions. They are not, however, free to set prices and have consequently accumulated significant losses and debt (Kong 2011).

In another reshuffle in 2003, the SETC was then abolished. The State Council established an Energy Bureau under the NDRC, which, being underfinanced and understaffed, has remained unable to coordinate the energy plans of firms and ministries. State-owned energy companies continue to be the most influential players within China's energy policy. They are actively involved in the drafting of energy laws and have occupied key positions within supervisory bodies (Downs 2008; Meidan 2016). In 2008, renewed attempts were made to create an effective national-level regulatory institution: the National Energy Administration (NEA). The NEA still exists at present. Like its predecessors, it has been equipped with a broad mandate but remains unable to mitigate the infighting among state agencies or to impose its will on China's largest energy companies. Another attempt at overall coordination was made with the establishment of the National Energy Commission (NEC) in 2010. The NEC is directed by the prime minister, and composed of 21 ministers and directors from different government bodies – including the Central Bank, the NEA and the NDRC (Kong 2011). Responsibility for important aspects of energy policy – especially pricing – still rests with the NDRC, which has repeatedly tried to obstruct the formation of a new energy ministry.

Electricity transmission and distribution are controlled by the state via corresponding licences issued and standards set by the NDRC. In 2002 the State Electricity Regulatory Commission (SERC) was established, charged with overall regulation. SERC also advised the NDRC on tariffs, but was then dissolved in 2013. Its functions have since been transferred to the NEA (Wilson, Yang and Kuang 2015). Also in 2002, the monopoly of the State Power Corporation was dismantled into separate generation, transmission and service units so as to improve overall system efficiency and introduce competition by creating wholesale markets – allowing direct purchasing by large industrial customers, and encouraging interprovincial trading. Retail companies have been permitted since 2015, meanwhile, but electricity generation is still dominated by five state-owned holding companies, with some electricity being generated by independent power producers, often in partnership with de facto state-owned companies.

Private and foreign investment in generation and transmission has been possible since 1997. This liberalisation initially generated lots of interest and optimism, leading to around 100 memoranda of understanding on electricity-related investment projects in the late 1990s. Still, a lack of confidence in China's continued growth, misgivings about power-purchase agreements with local counterparts – whose terms worsened for investors in 2002 – and a waning Chinese need for capital, management skills and technologies from abroad led to an eventual significant loss of interest on both sides. There are no exact figures available on the share of private investment in China's electricity sector today (Wilson, Yang and Kuang 2015; Xuegong, Lyian and Zheng 2012).

China's oil and gas sector is better structured than its coal one. Three state-owned enterprises that emerged out of a government-led reorganisation process dominate the former: the China National Offshore Oil Corporation (CNOOC), the China National Petroleum Corporation (CNPC) and the China Petroleum and Chemical Corporation (Sinopec). The CNPC is the largest and most influential of the three, followed by Sinopec. Both are integrated oil companies, meaning that their activities cover the entire value chain from exploration and extraction to refining, distribution and marketing.[5] The CNOOC, meanwhile, has limited its activities mostly to offshore exploration.

There are only a few and rather small other private and state-owned oil companies. Foreign firms have been granted access to offshore sites, mainly through production-sharing contracts and joint ventures, while onshore production is limited to the big state-owned companies. These command a vast workforce of almost 1.5 million people, according to information available on their respective websites, and contribute a significant share of the state's tax revenues. Top managers of these companies hold ministerial rank, giving them direct access to central government. Several current members of the Central Committee of the Communist Party began their career in the oil industry, which has given rise to a "petroleum faction" within the state's inner leadership circle (Williams 2014).

The political influence of energy companies, and especially the three giant oil firms, is revealed in the interviews conducted by Williams (2014) with government officials. Her interviewees maintain that oil firms delay the central government's energy-intensity programmes and resist the addition of renewable capacities to the grid. In recent years, these companies have become largely independent from political oversight. Most demonstratively, oil firms have been able to pursue their own foreign economic policies directly or through subsidiaries, initiating deals and overseas investments at times and in places when/where this was otherwise discouraged by the political leadership (Baumgartner and Godehardt 2012; Meidan 2016; Müller-Kraenner 2008). Probably the best example of this is investment in Sudan in 2007, when the African country was dropped from a list of states to do business with by the NDRC because of its poor international reputation – a result of its extensive human rights abuses (Kennedy 2010).

Seemingly, energy firms – even state-owned ones – are not driven by political considerations. At least in the case of China, going abroad instead of investing in domestic exploration and production is the preferred strategy because it allows these firms to expand their respective market shares globally, to diversify transport routes and to reduce market risks. Investment abroad also frees these enterprises from the pressures of excess labour, and makes them largely independent from political control (Downs 2008). The experience of Sudan in particular has made the political leadership attempt to discipline the energy companies, though. In 2011, their top executives were reshuffled. The companies themselves were forced to increase dividend payments to the central government. After President Xi Jinping came to power in 2012, he started an anti-corruption campaign whose most prominent casualties have been the patron of the oil sector, Zhou Yongkang, and his entourage. Opening up the oil and gas sector to private investment has been another means to break the power of established companies (Andrews-Speed et al. 2014).

Besides the strong position of energy firms, pursuing a coherent energy policy is hampered also by the considerable amount of autonomy enjoyed by the provinces in economic and energy legislation. This is exemplified by their resistance to the cessation of preferential electricity pricing for energy-intensive companies residing within their geographical limits, which has forced the central government to stiffen directives on a number of occasions (Zhou, Levine and Price 2010). The decentralisation of fiscal and tax policies in 1994 created incentives for local governments to look for new ways to boost their income from taxation. Local governments therefore often overshoot the centrally determined energy targets if producing more than planned yields local tax benefits. A case in point is that quantities of coal extracted in certain provinces are frequently higher than the amounts stipulated by the central government itself. Conversely, inefficient and highly polluting production facilities tend to be allowed to continue operating, as closing them down would decrease locally paid taxes and generate economic problems that the provincial authorities would have to deal with (Kong 2011; Williams 2014). Affirming these problematic tendencies, Van Aken and Lewis (2015) show that those provinces with more autonomy and a higher share of industry in GDP and tax income implement measures to save energy and reduce emissions more reluctantly.

An environmentally sound energy policy is, furthermore, obstructed also by pricing issues. Reforms in this domain have hitherto been piecemeal and insufficient. The government launched fuel tax reforms in 2008 so as to tie the retail price of oil and gas products more closely to those of international markets. Increases in domestic prices trailed behind international prices thereafter, however – causing refiners to incur losses (EIA 2015d). Coal prices are state controlled, with end-user prices having been raised for energy-intensive sectors by 50 to 100 per cent in 2010. Electricity prices for the residential sector, meanwhile, remain low at about half of the average of the Organisation for Economic Cooperation and Development (Ma 2015). The reason for this is that electricity prices still serve as a tool to support social stability. In addition the production of fossil fuels has been subsidised by budgetary support for state-owned energy companies, freight and tax discounts as well by government funding for corresponding research. Although limited public disclosure makes it difficult to assess the exact extent of these subsidies, they have been estimated to reach slightly more than USD 3 billion a year (Bath et al. 2015).

The expansion of hydropower and renewable energies has been hampered by certain political shortcomings. Large power providers are not permitted to include small hydro schemes in their legally mandated energy mix. Small hydro programmes suffer from low, locally regulated prices. Large dam projects have faced opposition from China's growing and increasingly assertive civil society, which succeeded, for instance, in bringing the Nu River Project – consisting of five additional large-scale hydropower stations along that river – to a halt in 2004 (Xie 2011). Regarding wind power, a sizable share of the installed capacity was never connected to the grid in a timely manner after commissioning because of inadequate coordination between developers and operators – which weighs heavily in a large country that suffers from a lack of long-distance grid connections (Wang, Yang and Zhang 2015). Legislation is being discussed to guarantee the purchase of all renewable energy output up to a certain amount. Beyond this limit, generators would be allowed to negotiate contracts with end users. These would be given priority dispatch rights.

The People's Republic – or rather the aforementioned oil firms – has also been incredibly active vis-à-vis secure scarce resources abroad. It is a key player in the scramble for the natural gas resources of Central Asia, fostering political ties with the post-Soviet republics there and building pipelines to gain access to the reserves of these countries. Already in the late 1990s, the considerable potential for regional cooperation on energy in Central Asia was highlighted – with China being a potential key customer that could, by building energy-related infrastructure there, also further develop its own province of Xinjiang (Dorian, Wigdortz and Gladney 1997; Dorian et al. 1999). China has indeed succeeded in upgrading energy infrastructure in Central Asia. Today, the Kazakhstan–China Pipeline runs all the way from Xinjiang to the Caspian Sea. It connects to the Russian pipeline network, too. The Central Asian Gas Pipeline, which links China and Turkmenistan, was completed within just 28 months, meanwhile.

Equity investment and long-term supply contracts, accompanied by concessionary grants or loans, have been other means to obtain energy resources from abroad. China's energy giants have concentrated their investments in regions where Western competitors have only a weak presence, beginning with Kazakhstan in 1997 (Günther 2005). Long-term contracts for the delivery of liquefied natural gas (LNG) have been concluded with firms from Australia, Indonesia, Malaysia, Papua New Guinea and Qatar. According to IEA (2014a) estimates, overseas oil and gas production by Chinese companies now stands at 2.5 million barrel of oil equivalent a day. Foreign production reaches 50 per cent of domestic production. In total, Chinese companies are active in 42 countries around the globe. The most important ones, measured by volume of investment, are listed in Table 1.1 below. Half of production – which does not correlate with the amount of money invested – comes from Africa and the Middle East. Twenty-one per cent is from Central Asia, 11 per cent from Latin America and 9 per cent from North America. Iraq has become the leading supplier – not for China's oil imports, which are, as noted, still mainly from Angola and Saudi Arabia, but for imports of oil extracted by Chinese companies. Kazakhstan and Nigeria are the second-most important countries in this regard.

Table 1.1 Chinese Overseas Oil and Gas Acquisitions from 2002 to 2013

Country	*Number of Acquisitions*	*Total Volume in USD billion*
Canada	11	31.0
Brazil	4	16.7
United States	8	13.0
Kazakhstan	7	11.1
Australia	9	10.5
Egypt	2	6.4
United Kingdom	4	5.5
Angola	3	4.8
Mozambique	1	4.1
Argentina	2	3.1
Syria	2	3.0
Peru	1	2.6
Colombia	3	2.3

Source: IEA (2014b).

Recently, there has been a certain shift to sites in Canada, United Kingdom and the United States, where Chinese companies have become active in mergers and acquisitions – destined for the development of North Sea blocks and unconventional resources in North America. These joint ventures are also meant to improve the managerial and technical know-how of the Chinese partners. Investments in developed countries, to which Australia has to be added, and in some parts of the Global South, in particular Brazil, are considered safer due to political stability, especially after very bad experiences with equity investments in Libya and Syria as well as the repercussions of the Western sanctions imposed on Iran (IEA 2014b).

The outcome from this massive investment in hydrocarbon resources abroad has been mixed, at best. From a political perspective, it is problematic that the majority of imports are shipped through seas controlled by the US navy, giving rise to fears about possible energy containment (Downs 2004). As the production costs of equity oil – including credits and grants for producer countries – appear to be higher than they are for contract oil, expert circles in China question whether the current strategy makes sense from an economic point of view (Kennedy 2010; Shaofeng 2011). Securing energy supplies abroad has also fuelled conflict with other import-dependent emerging economies, most notably India. Vis-à-vis India, China has had the upper hand in the competition for equity oil and long-term contracts (in Angola, Ecuador, Kazakhstan and Nigeria), mainly because these have been accompanied by concessionary financing. An agreement reached by China and India for joint development in third countries has brought little relief, as it remains bereft of any significant follow-through aside from certain minor endeavours undertaken in Indonesia, Iran, Nigeria and Syria (Gu 2011).

ENERGY-RELATED KNOW-HOW

Regarding the new normal, scientific innovation is stressed by the Chinese government as a key means not only to transform the domestic economy but also to shift towards a sustainable energy policy. Public expenditure on research and development has increased, now accounting for some 2.2 per cent of GDP.[6] Spending on energy research, especially for renewables, has grown rapidly during the last decade. Corresponding private and public investment in the People's Republic is as much as that in the European Union (EU) and United States combined. The bulk of this investment is public, and relates to solar energy (FS-UNEP and BNEF 2016). As a consequence, China has become the global leader in the production of photovoltaic cells, modules and wafers.

The solar industry has been declared strategic, and therefore benefits from low-cost loans and various types of subsidy. In 2012, Chinese companies account for 71 per cent of total global production of photovoltaic cells. Whereas the take-off of the industry was made possible by exports, the domestic market has been generating a considerable demand for about 10 years now. The shift towards the home market is particularly relevant because European regulatory agencies as well as the United States have accused Chinese companies of dumping, punished them with countervailing duties or have negotiated voluntary export restrictions with China itself (Puttaswamy and Sahil Ali 2015). Still, it appears that China's solar industry suffers from overcapacity – which has already led to several cases of bankruptcy (REN21 2015).

Wind-installation producers have seen a similar rise. In 2000, they delivered only a miniscule share of global installations – being produced by just four companies. By 2011, China had become home to 80 producers – 8 of which were among the top 15 worldwide. Contrary to solar power, wind-installation producers have supplied predominantly the domestic market from the very beginning. These companies have also been heavily involved in research activities. The People's Republic accounts for the second-most patent applications in wind power, still behind the United States but ahead of the EU. Given the size of China's domestic market, local content obligations imposed by the government on foreign investors and financial assistance provided to domestic firms have proven adequate means to support Chinese wind-installation producers. Most producers in China have by now become independent from foreign technology licences (Gosens and Lu 2014).

CCS – on whose general advantages and disadvantages we elaborate in chapter 7 – constitutes an important component of China's energy strategy because of the country's heavy reliance on coal, and resulting deterioration of air quality. With strong government support since the inception of the 12th Five-Year Plan, energy companies and research institutes have been conducting research on CCS, which has resulted in pilot and demonstration projects.

The first such pilot project – a combined heat and power plant – started operations in 2011. It is expected that China will have commercial-scale CCS demonstration projects by around 2020 (ENGO Network on CCS 2015). There has been significant international cooperation between China, the EU and the United States. Already in 2009 a Sino-American research centre was established, focussing on CCS and other technologies related to coal. Five years later, China and the United States jointly opened the Center for Excellence on Nuclear Security, located in Beijing.

The results of this latter endeavour have been rather modest: the centre's very limited budget of USD 150 million has been used to conduct studies and organise workshops; only a small number of patents have been generated (Yang 2016). An additional area of expertise is coal liquefaction, which was initially neglected after the discovery of relatively large domestic oil fields in the 1960s. In the early 1980s, interest in liquefaction techniques was revived as a result of China becoming a net oil importer (Sun 2010). The first and currently still only coal liquefaction facility – constructed with massive public support – reaches a capacity of 20,000 barrels a day, being second in size worldwide. It started operations in 2010. The construction of additional facilities has been discouraged meanwhile, mainly for environmental reasons (Rong and Victor 2011).

Nuclear fusion,[7] arguably set to be ready for commercial use in China in 50 to 60 years' time (*South China Morning Post*, 17 July 2017), is probably the most important area of energy-related research in the People's Republic – at least from a long-term perspective. Corresponding research is carried out at the Experimental Advanced Superconducting Tokamak facility by the Chinese Academy of Science. Chinese firms also participate in the International Thermonuclear Experimental Reactor (ITER), which is – measured by current budget – the single-largest international scientific cooperation project anywhere in the world. The People's Republic initially contributed USD 1.6 billion to ITER, which equals 10 per cent of the project's costs. ITER is expected to be ready for preliminary experiments by 2025. A Sino-French research centre on nuclear fusion was established in the Chinese province of Anhui in 2018. The so-called China Fusion Engineering Test Reactor is expected to start operating in 2035, and may first generate electricity only six years thereafter (*South China Morning Post*, 26 December 2017).

CLIMATE POLICY

China used to be a major roadblock to collective action against greenhouse gas emissions, repeating the mantra of common but differentiated responsibilities – which the government interpreted as an exemption for the country from binding commitments on the international level. This stands in a stark contrast to its implementation of serious energy-saving and emission-reduction programmes, as described above. The People's Republic has implemented more encompassing reforms than any other country dealt with in this book. It has also been more successful in reaching its own official targets for energy saving and emissions reduction. China was among the first states to use internationally recommended market mechanisms to support a shift in energy generation. The 12th Five-Year Plan stipulated the goal of gradually establishing a carbon-trade market, although the exact details of this were left open. In 2011, the NDRC ordered seven cities and provinces to set up pilot schemes for carbon trading. These were then implemented in 2013 and 2014. Carbon prices in the schemes, however, were set at a very low level of USD 8 a tonne of carbon dioxide equivalent at a maximum,[8] that compared to more than USD 50 a tonne in the Scandinavian countries (World Bank 2016). China also intends to further tighten fuel-efficiency standards and to promote electric vehicles. The People's Republic share in the global production of electric vehicles stands at an impressive 40 per cent. The production target for 2020 is 5 million such vehicles.

The interesting question vis-à-vis China is why environmental activism at the domestic level has not translated into a proactive international stand. It is plausible to assume that China wants to keep as much leeway as possible for future economic growth – which, regardless of the new normal, implies the continued growth also of energy consumption and greenhouse gas emissions. This potential economic growth would be restricted if the People's Republic signed binding international agreements, and even more so if it agreed to a reduction of energy-related emissions in absolute terms. Further to this, concerns regarding energy security and pressure from civil protests against environmental degradation – often related to the generation of non-renewable energy – have driven the policy adjustments witnessed in China. However these are only rather recent phenomena, and ones opposed by influential players such as the major oil companies. It appears to take more time for these new political considerations and forces to have an impact on the government's position in international negotiations.

The first instance of a narrowed gap between domestic activism and international reluctance was the UN Climate Conference in Copenhagen in 2009, where China put forward a relatively ambitious target of reducing the emission intensity of its GDP by 40 to 45 per cent by 2020 (compared to the 2005 level) and to increase the share of non-fossil fuels in primary energy consumption to around 15 per cent by the same year. In 2014, at the Asia-Pacific Economic Cooperation summit, the People's Republic promised in bilateral negotiations with the United States that carbon dioxide emissions would peak in 2030. It also committed to increasing the proportion of non-fossil fuels in its primary energy consumption to 20 per cent by that same year (*New York Times*, 11 November 2014).

According to China's INDC, the carbon intensity of its GDP will be reduced by 60 to 65 per cent by 2030 as compared to the 2005 level. The INDC reaffirms that the share of non-fossil energy will rise to 20 per cent. However, this leaves room for a considerable increase in energy consumption and carbon dioxide emissions in the coming years. Up until 2030, China's energy consumption is expected to increase by 81 to 89 per cent as compared to the 2010 level. Related emissions will rise by 68 to 74 per cent, meaning that the People's Republic alone will most likely absorb nearly 40 per cent of the remaining global carbon budget (Green and Stern 2015; Liu et al. 2015). The commitment made within the INDC is not overly ambitious. It can be achieved by still pursuing current policies, without the need to undertake any additional efforts; as such, it was initially rated "medium" by Climate Action Tracker. A more ambitious commitment – especially an earlier emission peak – would nevertheless have been possible (Green and Stern 2015).

Climate Action Tracker has by now downgraded China's commitment to "highly insufficient". The organisation acknowledges that there has been a relative reduction of the share of coal in the energy mix, and that targets for solar energy have indeed been overachieved. The expansion of wind energy is on track to exceed the 2020 target by a quarter. Climate Action Tracker also points out that China's carbon dioxide emissions appear to have already reached their peak. But this is far from a fair contribution to what would be necessary to limit global warming to less than 2°C, especially because emissions of other greenhouse gases continue to increase.[9] Whereas the other emerging economies are rated poorly by Climate Action Tracker because their energy policies barely contain measures that can significantly contribute to climate change mitigation, the implicit argument with regard to China appears to be that it could do much more than it is doing – namely, because of its tremendous energy consumption and related emissions (in absolute terms) on the one side and its rather good financial and technological capabilities on the other.

NOTES

1. This information is available online at: www.tps-data-portal.org/all-datasets.
2. The technical potential is a theoretical one, taking into account only natural parameters but not land use and other non-natural factors that hamper the installation of, for example, solar panels and wind turbines. It does not say anything about the economic feasibility of corresponding installations, either.
3. The terms "supercritical" and "ultra-supercritical" refer to the pressure in the steam generators of coal-fired power stations. Whereas bubbles can form in a

subcritical boiler, a supercritical steam generator operates at pressures above the critical one. Therefore liquid water immediately becomes steam, resulting in less fuel use and, as a consequence, fewer greenhouse gas emissions.

4. The tonne of oil equivalent – often also barrel of oil equivalent – is a unit of energy based on the approximate energy released by burning one tonne/barrel of crude oil. It is used by oil and gas companies as a way of combining oil and natural gas production and reserves into a single measure.

5. The oil and gas industry is divided into three sectors: The upstream one comprises searching for oil and gas fields, the drilling of exploratory wells and subsequently the drilling and operating of these wells proper. The midstream sector involves the transport, storage and wholesale marketing of crude or purified/refined products. The downstream sector is about refining crude oil and purifying raw natural gas. It also includes the marketing and distribution of products derived from oil and gas.

6. According data is available online at: www.oecd.org/sti/msti.htm.

7. Unlike a fission reaction, in which atoms collide and release energy, fusion is the result of particles being forced together at extremely high temperatures until they fuse into a heavier nucleus. This process generates vast amounts of energy. The relevant isotopes for fusion are less dangerous than those necessary for fission, and are therefore less likely to cause a widespread radiological incident.

8. The tonne of carbon dioxide equivalent is a measure used to compare the emissions of different greenhouse gases based on their global warming potential.

9. This information is available online at: http://climateactiontracker.org/countries/china.html.

Chapter 2

India

In spite of considerable potential for renewables, coal and oil play a dominant role in India's energy matrix – and, furthermore, they are increasingly being imported. The electricity sector is inefficient, also because of heavy subsidies. We also elaborate on the continued expansion of energy-generation capacities. Despite the recent shift towards natural gas and renewables in policy planning, this expansion will still lead to a dramatic increase in fossil fuel consumption.

Further to this, we shed light on inefficient institutional arrangements and the half-hearted liberalisation of the country's energy sector. India has, nonetheless, developed sound plans and policy instruments that possess the potential to affect energy generation and consumption in a positive way. We also summarise efforts to acquire electricity and fossil fuels abroad, both through worldwide investment and via cooperation with neighbouring countries. Concerning energy-related know-how, this chapter shows that India has a particular strength in nuclear technology. Research on solar and wind energy – and related production – is at only the initial stages. We argue, moreover, that the government retains as much leeway as possible in negotiations on climate change mitigation. The planned expansion of renewables is a major contribution in this regard though, arguably meaning that India will do its fair share to meet global targets.

MATERIAL CONDITIONS

India's energy sector is based on hydrocarbons. Coal and peat account for around 46 per cent of total energy supply, followed by biomass and waste (23 per cent), and oil (22 per cent). Natural gas (5 per cent), hydropower and renewables (3 per cent) as well as nuclear power (1 per cent) play only marginal roles. Around 35 per cent of India's energy requirements have to be imported – far more than in 1990, when this figure stood at 11 per cent. Import dependence is most pronounced for oil (83 per cent), less so for natural gas (36 per cent) but remains considerable for coal (50 per cent). Especially coal imports increased up until recently. For oil and gas ones, India is dependent on a handful of suppliers (Kuwait, Nigeria and Saudi Arabia) (EIA 2016a; IEA 2015a). These suffer from questionable economic and political stability. This has led Indian oil companies, supported by the government, to purchase equity stakes in overseas oil and gas fields (more on this later).

India has a total installed electricity generation capacity of 304 GW, 76 GW of which are owned by the central state, 102 GW by the federal states and 126 GW by private entities (Central Electricity Authority 2016a). A perennial problem of the power sector is large distribution losses because of outdated transmission lines. There are also considerable commercial losses experienced due to insufficient billing, poor metering and theft. Counted as one, these losses amounted to 38 per cent of total generation in 2003. That figure came down to 23 per cent in 2014, which is still among the highest loss rates anywhere in the world (Swain 2016). Regarding sources of electricity, 62 per cent of total capacity is based on coal. Hydropower (14 per cent), other renewables (14 per cent) and natural gas (8 per cent) are relevant, whereas nuclear power (2 per cent) does not play a major role (EIA 2016a; IEA 2015a).

The largest nuclear power station – Kudankulam, with a capacity of 2 GW – first went online only in 2013. Some of the 132 coal-fired power stations

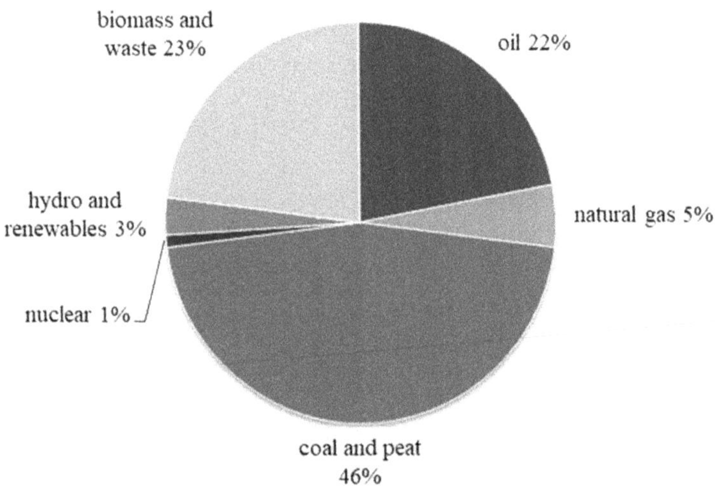

Figure 2.1 India's Primary Energy Supply by Source. *Source*: IEA data, available online at: www.iea.org/statistics.

in total reach a capacity of 2 GW or more, but the vast majority are rather small. On average, India's coal-fired power plants are among the least efficient in the world. This will change slowly with the successive installation of supercritical plants. Power stations fired with natural gas have mostly lower capacities, the exceptions being the Dabhol Station (2.2 GW) and the Samarlakota Plant (2.6 GW). Hydropower stations are also numerous and typically small. Only 10 reach a capacity of 1 GW or more. Solar parks and wind farms remain at individual maximum capacities of less than 1 GW.

In spite of rapidly increasing imports, India is the world's third-largest coal producer. It owns the fifth-largest coal reserves worldwide. There is some controversy about the volume of these reserves, traditionally assumed to be plentiful. Technically, only part of them can be extracted. At current levels of production, India's coal reserves will be exhausted in about 45 years' time (Camilleri 2011). More than 90 per cent is produced by open cast mining, which is cheaper than deep mining and thus partly explains why coal is so important to the country's energy generation. However, much of India's coal is of low quality because of its high ash content. It is therefore unsuitable for coking purposes. Another problem is that most reserves are located far away from the centres of energy consumption, being situated mainly in Chhattisgarh, Jharkhand and Telangana. Coal must be transported on average more than 500 kilometres to power plants. Adding to this problem is the inefficiency of the Indian railway system and its high freight shipping rates, which cross-subsidise commuter transport (IEA 2015a).

India has therefore had to increase coal imports over the last decade. This trend was reinforced by capacity addition in the electricity sector and the fact that domestic extraction by state-owned Coal India trailed far behind targets set by the respective five-year plans (Anand and Prasad 2013; Gupta, Shankar and Joshi 2012). Since 2013 there has been a remarkable upswing in domestic coal production, causing a dip in imports by more than 30 per cent in 2015 (Buckley and Sharda 2016). The recent decline of imports has been reinforced by the fact that demand has been sluggish since 2012 because of a certain degree of economic deceleration. New mines have been opened rapidly, especially after the fast-tracking of environmental clearances by the Modi government. Now, there is an oversupply of coal. The government has consequently slashed some of its expansion plans, as we summarise below.

India disposes of around 5.7 billion barrels of proven oil resources, a minuscule share of global resources: namely, 0.34 per cent. Production has stagnated since the mid-1990s because of maturing fields, only a few new discoveries and a lack of foreign investment. With rapidly increasing consumption, the share of imports has increased – making India the world's fourth-largest net importer of oil. Its supply gap has reached 4.1 million barrels a day presently. This contributes heavily to the country's import bill and balance of payments deficit. The EIA (2016a) expects that the share of imports will increase from 83 per cent today to nearly 90 per cent by 2040. India's own domestic reserves are located onshore in Assam, Gujarat and Rajasthan, and offshore in the Bay of Bengal and close to Mumbai. Oil remains underexplored as a commodity in India. The reason for this is that rather complex legislations has hindered investment in deepwater exploration and extraction to date.

Reversing a former steady increase in natural gas production, one made possible by a number of significant findings, domestic production has been marked of late by a sharp downward trend since its peak in 2010. Since then, the share of imports has almost doubled, reaching the aforementioned 36 per cent. Imports of LNG, mainly from Qatar, first began in 2004. Two import terminals are operational by now; another two near completion. The bulk of India's own domestic natural gas production comes from offshore fields in the Bay of Bengal and close to Mumbai. It is complemented by onshore fields in Assam, Andhra Pradesh, Gujarat and Tamil Nadu, whose output is much lower.

The potential of unconventional natural gas from coalbed methane and shale gas is considerable. Commercial extraction is low, but nevertheless gaining momentum (EIA 2016a; IEA 2015a). Natural gas consumption also suffers from an inadequate transmission and distribution network. Pipeline density remains one of the lowest in the world (Ernst & Young 2011), although it has doubled since 2008. The eastern and southern parts of the country are still not interconnected. There are no cross-border pipelines that could facilitate imports, but a number of projects have been discussed at least (more on this later).

Resources of uranium are very modest (less than 2 per cent of global ones) as compared to the country's actual use of nuclear power and, even more so, future ambitions. They are mostly of low grade, located in remote areas and expensive to exploit. Renewable energy resources, meanwhile, are abundant and spread relatively evenly across the country, with the exception of hydropower. Among renewables, traditional biomass – historically being a cooking fuel – dominates, but it will lose importance with increasing rural electrification. The use of biomass for electricity generation stands at a total of 5 GW. Additional capacity that could be generated from this source is expected to reach 23 GW (EIA 2016a). The technical potential of solar power is estimated at 750 GW, 479 GW of which could be exploited economically by 2047 according to optimistic calculations. Solar irradiation is strongest in the north and west of India, where installed capacity has also been growing the fastest. Costs of solar power have come down rapidly since 2010, and may reach grid parity by 2040 (IEA 2015a; National Institute for Transforming India 2016).

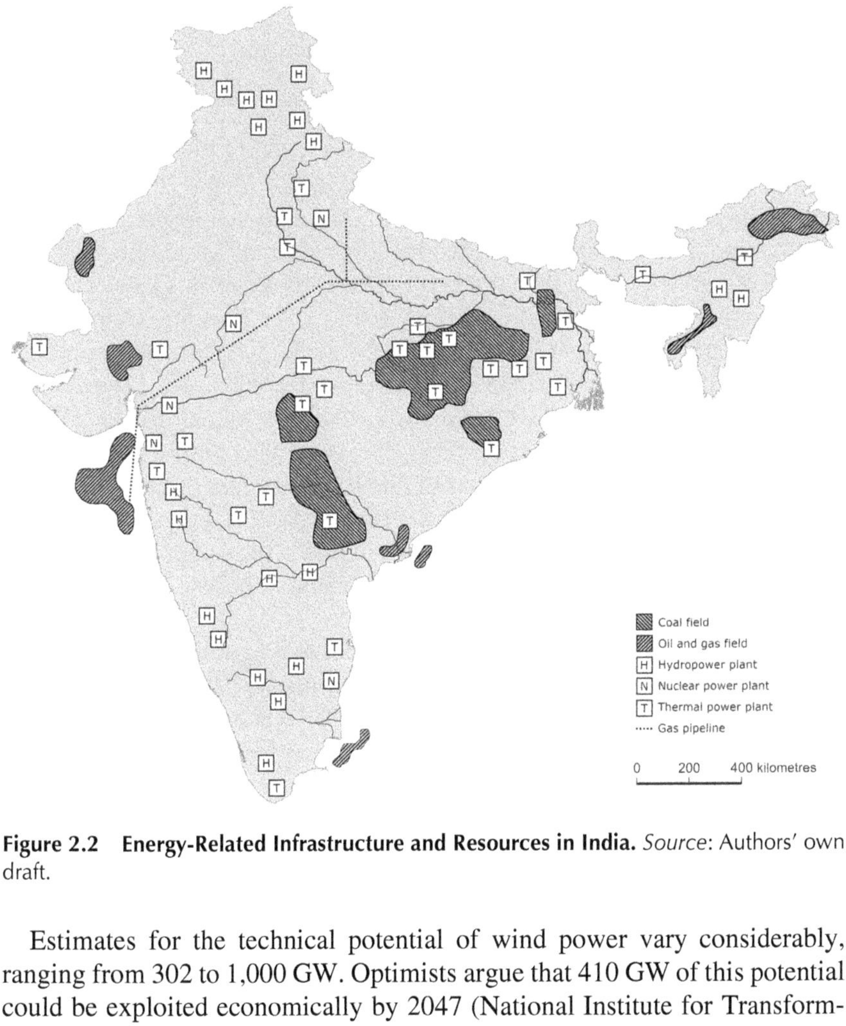

Figure 2.2 Energy-Related Infrastructure and Resources in India. *Source*: Authors' own draft.

Estimates for the technical potential of wind power vary considerably, ranging from 302 to 1,000 GW. Optimists argue that 410 GW of this potential could be exploited economically by 2047 (National Institute for Transforming India 2016; NITI Aayog 2015a). The best sites for wind farms are in the south and west of the country. The contribution of hydropower to energy supply is in relative decline, but the remaining technical potential is considerable: three-quarters of the total potential are untapped, equivalent to 149 GW. The overwhelming majority of the unused potential is located in the northeast and the Himalayan region (Planning Commission 2013). Sediments in rivers, however, pose problems, because they reduce dam capacity. In general, non-hydro renewables are hampered by their remote location, which means that potential sites are poorly connected to the actual centres of energy use. The cost of building the necessary transmission infrastructure would be tremendous, especially for private utilities that suffer from high costs of credit – which amount to about 12 per cent (Kaladharan 2016).

ENERGY DEMAND AND EXPANSION PLANS

India is already the third-largest producer of electricity in the world, showing the most rapid growth in demand worldwide – a trend that is only likely to continue (Buckley 2015; IEA 2015a). In 2013, its share in global electricity generation reached 4.8 per cent (EIA 2016a). The Planning Commission (2013) expects energy demand to increase by 5.4 to 5.7 per cent a year until 2022, and a quadrupling by 2035 as compared to the 2005 level. The successor institution of the Planning Commission – the Policy Commission, known in Hindi as NITI Aaayog (2015b) – predicts even higher annual growth rates of 5.7 to 7.4 per cent. It has to be taken into consideration, though, that India has not always met the targets set for the expansion of its energy generation. This shortfall was around half of capacity addition at the end of the 10th Five-Year Plan (2002–2007) and about 40 per cent at the end of the 11th Five-Year Plan (2007–2012).

In spite of these uncertainties, the IEA (2016a) argues that energy consumption in India will grow by 135 per cent between 2014 and 2040. India is, moreover, marked by the fastest growth of coal consumption worldwide (from 378 million tonnes of oil equivalent today to 936 million tonnes thereof in 2040). The share of coal in overall energy consumption will rise to 48 per cent by 2040. Per capita energy consumption will increase from a third of the world average to about 60 per cent. Because of this trend as well as continued population growth, India is expected to account for a quarter of global energy demand growth from now until 2040 – more than any other country worldwide. On account of the dominance of coal in its energy mix, this emerging economy will also be the country with the fastest rise in volume of greenhouse gases emitted (IEA 2015a).

Power-generating capacity increased during the 11th Five-Year Plan but, as noted, remained far below the stipulated target: 55 GW were added to the grid instead of 78.7 GW. The shortfall was largest in hydro and nuclear power and the state sector, whereas private players overachieved (Planning Commission 2013). The reasons for the overall underachievement were poor planning, delays on the part of equipment providers and unresolved land conflicts. Planning and construction improved and scaled up afterwards, which led to a rise by 50 per cent in capacity between 2012 and 2016 (Central Electricity Authority 2016b). However, despite the increase in power-generation capacity, India faces a shortage in electricity supply, causing frequent blackouts and forcing many firms to invest in diesel generators as a backup.

Given the aforementioned advantages of coal, it is not surprising that thermal power accounted for 58 per cent of planned capacity addition in the 11th Five-Year Plan. Hydropower was the second-most important source of electricity, with a share of 35 per cent. Nuclear power (6 per cent) and non-hydro renewables (1 per cent) played marginal roles. Among thermal power sources, a relative increase was foreseen for natural gas. Demand for this commodity has, indeed, risen tremendously in recent years, also for purposes other than electricity generation. Distribution infrastructure to private households has improved, end-user prices for small consumers are reasonable and gas-fired vehicles have become more widely used (Ebinger 2016). The share of coal in total generation capacities was, meanwhile, to remain flat, whereas oil should diminish in importance. This tendency was reinforced by the 12th Five-Year Plan (2012–2017): some 35 per cent of the additional capacities were to be based on natural gas, also reflecting the encouragement of households to switch to modern sources of cooking fuel and for car owners to instead run their vehicles with compressed natural gas so as to reduce smog. Non-hydro renewables accounted for another 25 per cent of the capacities to be built up in the course of the 12th Five-Year Plan, with solar (10 GW) and wind power (15 GW) contributing impressive amounts of electricity. In total, 80 to 100 GW of new electricity-generation capacities were foreseen by the 12th Five-Year Plan. The efficiency of existing thermal power plants was to increase. Half of the new thermal power stations envisaged by the 12th Five-Year Plan and its immediate predecessor were to be supercritical ones.

Additional electricity generation during the 12th Five-Year Plan exceeded targets by a slight margin in 2014/2015 but fell moderately below them the year after. The latest predictions available at the time of writing suggest that capacity addition will eventually be almost 15 per cent above target levels. Hydropower and nuclear power trailed far below these, however. Thermal power overachieved. With regard to renewables, the planned capacity addition will be overachieved – but only by solar energy, not by other sources. Private entities were much more successful in expanding their generation capacities than state-owned electricity providers were (Government of India 2016).

The newly formed Policy Commission has not made precise forecasts for the years to come, but in the Draft National Energy Policy it has indicated possible outcomes at least – ranging from business as usual to an ambitious agenda (without defining what "ambitious" actually means). Within this range, energy demand is projected to increase 2.7 to 3.2 times between 2012 and 2040. Electricity use will increase by a factor of 4.5. Energy consumption per capita will more than double. A massive increase of fossil fuel consumption is expected, therefore. Coal-based power capacity will increase to 330 or even 441 GW by 2040. Coal mining will double and eventually reach a plateau of 1.2 to 1.3 billion tonnes in 2037. Oil is expected to remain at its share of around 26 per cent in energy consumption, while that of natural gas will increase to 8 or 9 per cent by 2040. The Draft National Energy Policy recapitulates the renewables target set for India's INDC at the UN Climate Summit in Paris (more on this later). By 2040, a renewables capacity of 597 to 710 GW is predicted (NITI Aayog 2017). However, it remains to be seen whether the various distribution companies have sufficient financial capacities to buy as much electricity as they should. Non-compliance with renewable energy obligations is still widespread.

The targets for the nuclear sector are equally ambitious: capacity should increase tenfold already up until 2030, reaching 63 GW. This would require the construction of at least 30 new reactors (Ebinger 2016; Mohan 2016), which is unlikely in such a short time frame at least considering experienced delays in the past. Conflicts over liability in the case of nuclear accidents remain unresolved, especially with potential operators and suppliers from France and the United States. The selected reactor types have never been operated before. Cost overruns are enormous, and the French Nuclear Safety Authority has threatened to deny the certification of the reactor design that is destined to be applied in India (Ebinger 2016). An Indo–Russian announcement from 2014 on the construction of at least 10 nuclear reactors in the course of the next two decades has not led to concrete results as of yet (World Nuclear Association 2018). Opposition to new uranium mines and the construction of nuclear power plants has also been mounting for some time now (Ramana 2010). Given the aforementioned shortage of domestic uranium resources, the government intends to make use of thorium instead. This metal is available in large quantities: India is home to 25 per cent of the world's corresponding reserves. A pilot thorium reactor is scheduled to be in operation by 2022. Commercial use is expected by around 2030 (Mohan 2016).

To address the problem of technical electricity losses, the government launched the Accelerated Power Development and Reform Programme (APDRP) in 2002. It was meant to finance the modernisation of transmission and distribution networks. With the help of this programme, technical losses should be reduced to 15 per cent. Yet the success of the APDRP would remain only limited, as grants were provided to the federal states without obliging them to undertake concrete efforts to upgrade the transmission and distribution infrastructure (Khurana and Banerjee 2015; Planning Commission 2011). Therefore, the APDRP has since been restructured on the basis of demonstrated performance. The outcome of this new programme has also been rather disappointing: financial means have been far below anticipated targets, a significant diversion of funds has occurred, technical and commercial losses have scarcely been reduced while the objective of 100 per cent metering has also not been met (Ministry of Power 2016).

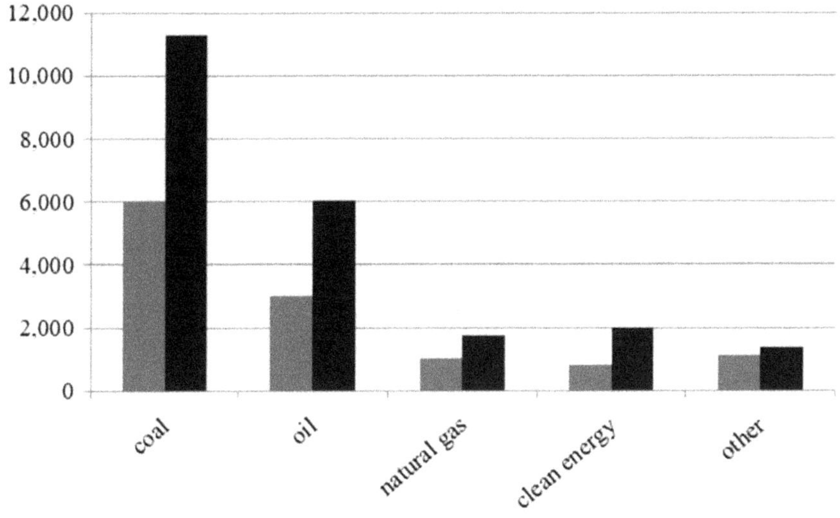

Figure 2.3 Expansion of India's Electricity Generation Capacities, terawatt-hours (TWh) (business as usual). *Source*: NITI Aayog (2017). *Note*: The years of reference are 2022 (light grey bars) and 2040 (dark grey bars). The Indian authorities subsume CCS, hydropower, nuclear energy and renewables under the term "clean energy".

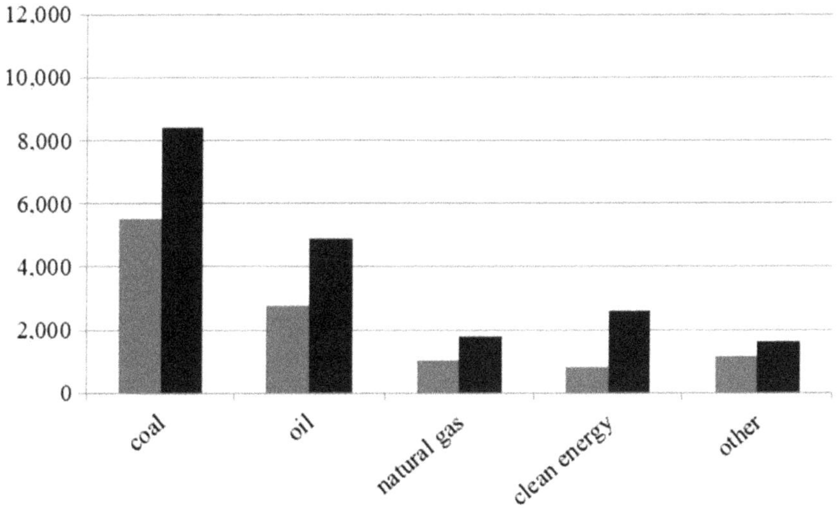

Figure 2.4 Expansion of India's Electricity Generation Capacities, in TWh (ambitious). *Source*: NITI Aayog (2017). *Note*: The years of reference are 2022 (light grey bars) and 2040 (dark grey bars).

Another major challenge for India's energy policy – at least with regard to climate change mitigation – is the envisaged expansion of its coal sector. As noted, coal mining is to increase. Numerous new coal power plants are planned, or already under construction. It appears that market mechanisms somewhat work against the oversized expansion of coal mining and coal-based power generation. As already mentioned, coal production increased too greatly in the recent past. The government had to revoke expansion targets for mining because of mounting stocks. Coal India was instructed to produce only as much coal as the economy could absorb. Declining utilisation rates of coal power plants – dropping from 74 per cent in 2011/2012 to 61 per cent in 2014/2015 – will probably lead to further stranded assets, in particular if the sector expands again – as envisaged in the Draft National Energy Policy (NITI Aayog 2017).

Large-scale coal projects with individual capacities of 4 GW each, originally announced in 2005, have run into difficulties of late because of delays ensuing from problematic land acquisitions – only adding to their high capital costs. Out of a total of 16 planned plants, only two have actually become operational (Sharda and Buckley 2016). One in Maharashtra has been closed, indeed, because of water shortages, a problem that will only become more severe in the future (Buckley 2015). The rapid deployment of renewable energy and their falling unit costs have led to strong competition for fossil fuel-derived power. It now appears that at least the immediate expansion of coal-fired power generation has been cancelled: the government has announced that no new plants will be needed between 2017 and 2022. Expansion plans for 2022 to 2027 will be upheld, however. Beyond that, a number of old and inefficient coal-fired power stations with a combined capacity of 5.5 GW have been identified for impending retirement.

Slowdown is not endemic to the environmentally most harmful sources of energy, though. The expansion of the natural gas sector depends on the global oil price, the willingness of domestic consumers to pay more for a source of energy cleaner than coal and the reduction of subsidies for fertilisers – whose production absorbs a large part of India's natural gas supply (Ebinger 2016). Of late, a large share of installed natural gas power capacity has lain idle because of a lack of available fuel at a reasonable price. The one for natural gas proscribed by the government is an insufficient incentive for exploration and development of domestic resources (Buckley 2015). Imports, meanwhile, are hampered by the aforementioned absence of international pipelines and the shortage of LNG terminals. India also faces difficulties to secure enough imports because of competition from other customers, as well as due to price conflicts with suppliers such as Qatar.

POLITICO-ECONOMIC CONSTELLATIONS

India's energy policy suffers from administrative inefficiency, overlapping responsibilities of public institutions and from a lack of coordination among the latter. At the national level, five ministries are directly involved in energy policy, complemented since 2005 by the Energy Coordination Committee – created so as to achieve a minimum level of policy consistency. Whereas the central government has exclusive competence for mineral resources, oil and nuclear energy, the federal states have jurisdiction over a variety of important issues, including natural gas infrastructure for example. They also tax mining rights and the sale and use of electricity. The model of cooperative federalism that the Modi government promotes implies an even larger degree of devolution in energy policy, which can be expected to soon reinforce related shortcomings, too.

Although institutional arrangements are, hence, problem-laden, reforms of the energy sector actually started earlier than in many of the other emerging economies. They initially centred on the deregulation and privatisation of energy production, aiming at dynamism through competition. The New Exploration Licensing Policy (NELP), formulated in 1997 and in force since 1999, has allowed for the engagement of private firms in the exploration and exploitation of oil and gas. It was passed because the government had by then realised that state-owned companies did not possess the technologies necessary to explore deepwater deposits or, indeed, the financial muscle to acquire such technologies. The Electricity Act of 2003 abolished the licencing system for power generation, enabling private companies to freely generate, transmit and sell electricity to end users. Other legislation from the early years of the new century aims at increasing energy efficiency: the Energy Conservation Act has empowered the government to prescribe energy consumption norms and standards for commercial buildings and energy-intensive industries. A Bureau of Energy Efficiency has been put in charge of developing consumption standards for appliances and buildings, too (Betz 2012).

A few years later, the government began supporting renewable energies alongside formulating more specific policies on energy saving. The Integrated Energy Policy was adopted in 2006, seeking to promote energy efficiency in all sectors of the economy and also the production of biofuels. This policy was also meant to accelerate the development of hydro and nuclear power. It has led to an increase of public funding for research on clean energy technologies (more on this later). As part of the National Action Plan on Climate Change of 2007, eight so-called national missions have been formed. They deal with sustainable agriculture, energy efficiency, "green India", sustainable habitats, the Himalaya ecosystem, knowledge on climate change, solar energy and water (Council on Climate Change 2007). The mission on solar energy has played a prominent role in scaling up the targets for the deployment of this form of energy generation. With the help of fiscal incentives, capital subsidies and a special tariff surcharge, solar power is set to achieve grid parity by 2022 (Parikh 2012). Within the mission on energy efficiency, a detailed plan of low-carbon strategies for energy-intensive appliances and industrial sectors has been developed. This mission has also advanced plans on how to boost non-motorised and public transport in cities.

What is more, large power utilities have been obliged to purchase renewable energy since 2003. Since 2010, the corresponding acquisitions have had to be verified by renewable energy certificates (RECs). RECs are available to accredited generators, and can be sold at prevailing market prices. Corresponding targets are set by public electricity regulatory commissions. Participation currently remains rather low though, and only a handful of federal states have enforced obligations and penalties for non-compliance to date. Consequently, long-term stable markets for RECs have not formed yet (Shrimali, Tirumalachetty and Nelson 2012).

The development of solar and wind power are also supported by tax incentives, mandatory feed-in regulations and preferential tariffs. However, the planned wider deployment of renewables – including hydropower – faces further obstacles. The expansion of large-scale hydropower is hampered by the difficulty of land acquisition, popular protests against new dams and, as a consequence, delays to many projects. The exploitation of small-scale hydropower is problematic because of high capital costs (as compared to larger projects) and insufficient connections to the national grid (Mittra 2011). The expansion of wind power is constrained by time-consuming land acquisition and approval procedures as well as by the elimination of incentives in 2012, as a result of fraudulent schemes (Kaladharan 2016).

The coal sector is dominated by state-owned companies. Only they are allowed to own and operate mines for general supply, meaning for sale on the Indian market. Coal India accounts for slightly more than 80 per cent of domestic production and is the second-largest employer in the country, with a total of 370,000 workers. Its operations are far less efficient than those of other coal companies because of the overly large workforce, strong unionisation and the persistent sheltering from competition (Ebinger 2016; Madan 2016). The remaining share of coal production comes from federal-state companies and private firms. Output by the latter is used exclusively for internal consumption. The overseas stake in coal mining remains very small, even though the sector has been open to foreign investment since 1993. Foreign investors may take up to 50 per cent of equity in joint ventures, producing for the respective purposes of these (IEA 2007).

Hence, foreign investors are likely to remain cautious because they have not been granted the right to sell the output of their mines. Apart from restricted market access, a major reason for the limited success in attracting foreign investment is that Coal India still plays a key role in identifying hurdles to private participation and it appears that the near monopolist retains the best sites for itself (Group Centennial 2009). Another obstacle is the requirement to transfer 26 per cent of mining revenues to affected communities, the failure of which has often caused unrest (Ebinger 2016). A proposal to allow private investment in coal mining for the general market was pending from 2000 up until 2015, mostly because of resistance thereto from labour unions (Ahn and Graczyk 2012; Buckley 2015). Even though this proposal has since been reintroduced to the policymaking process, it has not been passed by parliament and there are no indications of this happening any time soon either.

The expansion of coal mining and coal power projects have also been slowed down by growing popular resistance to new sites and the concomitant delay in environmental clearances and land acquisition. Another grave problem was corruption in the granting of mining licences under the Singh government. The vast majority of licences were granted at a heavy discount to politically connected firms and individuals, often with only limited expertise in coal mining. Most of the licences were eventually annulled by the Supreme Court in 2014 and handed back to Coal India (*Financial Times*, 24 September 2014). The Modi government would auction them again from late 2015 onwards. It has also made a determined effort to clear the investment backlog, and to simplify or even scrap altogether environmental clearances.

Still, even under Modi, coal prices are set by public companies, meaning first of all by Coal India. Power generators benefit from more favourable rates than industrial consumers do. Compared to imported coal with the same calorific content, and taking transport costs also into consideration, this system has generated an implicit subsidy for local consumption that amounts to something between USD 2 to 10 billion a year (IEA 2015a) – obviously thereby making it harder for other energy sources to compete.

In spite of the aforementioned NELP, India's oil production has remained flat for more than a decade now. As noted, this is mainly due to the complex regulatory environment for investment and, until recently, difficulties in raising oil prices to world market parity. Although there are no formal obstacles to private investment in the upstream sector, corresponding activities are dominated by Oil India and the ONGC – together accounting for two-thirds of total production. Oil India and the ONGC also account for the bulk of natural gas production. Since the late 1990s, more than 300 blocks have been offered up for exploration in auctions and more than 100 discoveries have been made. Yet, production has started in only a few blocks. The interest of foreign investors remains subdued because India's down- and midstream sectors are dominated by state-owned utilities. The Indian Oil Corporation operates 10 of India's 18 refineries, and controls around three-quarters of the domestic pipeline network. Retail prices of diesel, kerosene, domestically produced LNG and petrol were controlled by the government up until the beginning of this decade, significantly decreasing the prices paid by refineries for domestic crude oil (EIA 2016a).

The discrepancy between domestic and foreign prices for crude oil used to be made up by producers, refineries and the government itself through bonds and cash. With escalating oil prices in the first decade of this century, this system became unviable however. The government announced the deregulation of diesel and petrol prices in 2010. This policy became effective in consecutive steps from 2012 and 2013 onwards respectively. Only kerosene subsidies have remained in place, reflecting the importance of it as a cooking fuel for the poorest strata of society. With regard to natural gas, deregulation has also affected the midstream sector. The Gas Authority of India held an effective monopoly on transmission and distribution up until 2006, when new legislation first allowed private investors to hold 100 per cent equity stakes in pipeline projects. Since 2013, careful steps have been taken to adjust prices to international levels. Previously, natural gas prices were regulated according to supply sources, the customer in question and involved federal states (EIA 2016a). Natural gas cannot be sold on a purely commercial basis. It is allocated to different consumers by the government. Priority is given to fertiliser production and power generation. A new pricing scheme was approved in 2013, in an effort to attract private investment so as to increase domestic production and slow down the increasing dependence on imports. The new formula linked domestic prices to a basket of international prices, but it did not increase them enough to provide a sufficient incentive for investment (IEA 2015a).

Electricity generation has also been liberalised only half-heartedly. Since the adoption of the Electricity Act in 2003, increasing competition has been a major objective pursued by the government. Large-scale end users are to be given a choice of their respective suppliers, made possible by open access to the transmission and distribution systems. The federal states, however, have be in no haste to implement the corresponding reforms. Nearly all of them have passed regulation for open access, but difficulties remain for electricity providers because of high transmission charges issued by the state-owned operators of the power grid or – in some cases – outright denial of access. Some federal states have restricted power generators from supplying customers in other federal states and from obtaining electricity generated beyond the borders of the respective home state. The reason for these restrictions is that the federal states are seeking to protect the revenues of their own distribution companies in this way (Singh 2016).

Hence, a key obstacle to introducing market principles and related dynamics to electricity generation is that transmission and distribution are still dominated by public sector companies and licenced dealers. Most new companies remain under the holding structure of state electricity boards, meaning that prices and regulations are set by the governments of the respective federal states – who are usually unwilling to pass on rising costs to customers. Newly established regulatory commissions have been given limited jurisdiction. The average tariff for agriculture stood at INR 2.6 per kilowatt-hour (kWh) in 2014, compared to an average commercial tariff of INR 6.3 per kWh. Tariffs have been raised infrequently and insufficiently for fears of electoral backlash, which explains why cost recovery was only 82 per cent in the period from 2003 to 2011. As a consequence, transmission and distribution companies are unable to invest in additional transmission lines, in the modernisation and upkeep of the power grid, and quite often also to buy the required amount of electricity from the power generators – causing intermittent blackouts, as well as resistance to more expensive renewable energy sources (Khurana and Banerjee 2015).

The quest for energy security has transformed India's foreign policy in a similar manner to how it has that of China too. India is vying for a share in the exploitation of oil and gas reserves not yet monopolised. For this purpose it has deployed a new kind of energy diplomacy, accompanied by the build-up of naval forces to secure newly won assets and supplies (Müller-Kraenner 2008). In this race for secure supplies, India has most often lost out to China. The People's Republic first outbid India in Angola and Nigeria, then in Ecuador and Kazakhstan – and, more disturbingly for India, also in the near abroad, in Bangladesh and Myanmar to be precise. As noted in chapter 1, China and India have signed an agreement for the common exploration and development of oil and gas in third countries (Chatterjee and Sinha 2006). This agreement has led to very few common activities, mostly joint exploration – for example in Myanmar, Sudan and Syria. Following another bilateral agreement signed in 2012, common infrastructure projects such as a pipeline from Myanmar to south-west China have been realised; so have jointly operated production sites in Sudan and Syria (Reischer 2012).

Despite the failure to outcompete China, Indian companies held assets in some 24 countries in 2011 worth around USD 13 billion – mostly as minority owners with small shares. These assets were spread globally, including in Australia, Azerbaijan, Brazil, Colombia, Gabon, Indonesia, Iran, Iraq, Kazakhstan, Libya, Mozambique, Sudan, Syria, the United States, Venezuela and Vietnam (PwC 2015). The Modi government has encouraged state-owned oil firms to acquire more assets abroad at a time of low prices. For this purpose, it has inked confidentiality agreements for half a dozen new opportunities, the most prominent being a stake in Russia's Rosneft. Still, one should not overestimate the impact of equity oil and gas on energy security. Production overseas was 10 per cent of domestic demand in 2011, equalling 2 per cent of total energy consumption (Planning Commission 2012). In view of the inability of domestic coal companies to supply the required amount of resources, the government also encouraged the acquisition of equity in foreign coal mines and companies up until 2014. Most of these ventures – specifically, in Indonesia, Mozambique and the United States – lost money or even went bankrupt (Buckley 2015).

Regional cooperation – in particular in South Asia as well as with Iran and Myanmar – could contribute significantly to India's energy security, by providing electricity and natural gas. Yet corresponding projects remain few and far between, while it appears that interest in developing energy resources as part of regional cooperation is minimal at present (Wijayatunga, Chattopadhyay and Fernando 2015). Several projects are worthy of mention, despite being small scale, at only an early state of realisation and/or at a standstill:

- Bhutan supplies hydropower to India. The maximum transfer is 1,540 MW, which is of course only a fraction of India's demand.
- The Modi government, in pursuit of a more pronounced neighbourhood policy, has signed a preliminary pact with Bhutan for the construction of four hydropower stations. They are expected to deliver 2,120 MW, almost exclusively to be used by India (*Times of India*, 28 May 2014).
- Cooperation with Nepal, long stuck because of its mistrust of India's motives, has also advanced in recent years. In 2014, a USD 1 billion deal for a bilateral hydropower scheme was signed. This deal will deliver a 900 MW project, of which Nepal will receive 22 per cent for free and with the remainder being exported to India (*Daily Mail*, 25 November 2014).
- A pipeline from Myanmar via Bangladesh to India appears to have received its final death knell when Myanmar's junta decided in 2007 to move ahead with China as its partner instead (Macdonald 2008).
- A proposed pipeline for Iranian natural gas has been stuck in limbo because of disagreements between India and Pakistan on the transit fees and generally deteriorating Indian–Pakistani relations in the aftermath of the 2008 terrorist attacks in Mumbai. A private consortium from India is now advancing plans for an underwater pipeline that would bypass Pakistan and its territorial waters (*First Post*, 6 September 2017).
- The exploitation of vast natural gas reserves in Bangladesh has long been intensively discussed in both countries. Domestic opposition in Bangladesh to a supposed "sell-out" to India has brought the programme to naught so far. Trilateral cooperation with Bhutan as a provider of electricity

to Bangladesh – with India as a transit country and potential investor – appears more promising (*Dhaka Tribune*, 18 December 2017).

Aiming at access to more distant resources, the Indian government launched its Connect Central Asia Policy in 2012, signalling that it would foster economic and political relations with the states of that region. Transregional energy infrastructure is a key component of this strategy (Dave 2016; Gupta 2013). The ONGC has invested in Kazakhstan, but there appears to be no clear strategy on how to get Central Asian resources to India's domestic market. Because of the Taliban insurgency in Afghanistan and political friction with Pakistan, India is cut off from Central Asia. This problem is best exemplified by the Trans-Afghanistan Pipeline, which is to connect Turkmenistan to India. Ever since a first project started in 1995, there have been repeated setbacks. The pipeline is now expected to be inaugurated before 2020 (*Times of India*, 23 February 2018).

ENERGY-RELATED KNOW-HOW

India's total spending on research and development – including both private and public sources – is not particularly large, reaching merely 0.85 per cent of the GDP – although it has been growing faster than the economy for a while now. An impressive share of 22 per cent of total spending is destined for nuclear energy. Renewable energies receive 0.7 per cent hereof. Nuclear power has always been a public priority, a tradition dating back to the early 1950s. For decades, the civil nuclear programme had to proceed without nuclear fuel and technological assistance from other countries because of India's nuclear armament and abstention from the non-proliferation treaty. India was able to develop substitutes, and today masters the full nuclear cycle. After signing an agreement with the United States in 2008 – which put some of India's reactors under surveillance by the International Atomic Energy Agency but in principle acknowledged the country as a nuclear power – sanctions were lifted and agreements on civil nuclear power were concluded with several countries other than just the United States. The 2008 agreement hence paved the way for foreign suppliers to deliver nuclear fuel, reactors and technologies. Investments were held up, however, by India's nuclear liability law, which made suppliers responsible for accidents (unique in international practice). This problem has now been dealt with by the creation of an insurance pool for operators and suppliers (IEA 2015a).

As noted earlier, thorium as a fuel for nuclear power stations is to play a critical role in India. According to optimistic expectations, as much as

30 per cent of the electricity demand will be met through thorium-based power generation by 2050. The Atomic Energy Commission argues that thorium is the way to reduce dependence on (imported) fossil fuels and contribute to climate change mitigation. It also praises the supposed safety of reactors fuelled by thorium, and suggests that these could be set up in highly populated regions such as the metropolitan areas of Delhi and Mumbai (*India Today*, 14 February 2014). Researchers worldwide have studied thorium-based fuel cycles for almost 70 years, but India leads the way when it comes to related commercialisation. In 2002, construction of a 500 MW prototype fast-breeder reactor was approved. It would be built in the city of Kalpakkam. The entire system was developed in India. Apart from running on thorium, this type of reactor extracts up to 70 per cent more energy than a conventional one does. In the next decade, six more fast-breeder reactors are set to be built. It has also been argued that the rather small Indian reactors are very suitable for exporting to countries with a low total electricity demand (*Guardian*, 1 November 2011).

Research on renewable energies is carried out overwhelmingly by public agencies and, as noted, received a boost with the ratification of the Integrated Energy Policy. Public expenditure increased to a still rather low volume of USD 120 million a year at the end of the 11th Five-Year Plan, while being expected to double during the 12th Five-Year Plan. This expenditure is almost entirely on solar technologies, because of the corresponding preference for that of the Singh government – and, even more so, of the Modi one, too (Goerild et al. 2014). Public research on energies other than nuclear and renewables is mostly about CCS, which is labelled a clean energy technology. The Department of Science and Technology has taken a lead role in this regard, setting up a programme on carbon sequestration in 2007. Around 30 related research projects have followed. The ONGC had planned a CCS demonstration plant on its own, but cancelled it for reasons of cost. The National Aluminium Company wants to set up a CCS unit at its coal-fired plant in Orissa. Private Indian companies showed initial interest in CCS. They appear to have since abandoned this technology however, because of regulatory hurdles and storage challenges (Global CCS Institute 2013). The latter obstacle – especially probable public opposition to underground storage sites – explains why the government has taken a rather cautious approach towards implementing CCS.

The large-scale development of wind power in India offers the potential to develop expertise that can be sold internationally, although this field is not a significant area of research for India. There is already evidence of this, with Suzlon – the largest Indian company – now operating factories in China and the United States. Nineteen wind turbine manufacturers produce types of up to 2 MW in India today. Their technological sophistication has improved, but they are still dependent on critical components originating from foreign suppliers. Overseas companies have bought Indian wind turbine manufacturers such as Continuum Wind Energy and Green India, calling into question whether India will become the host country of lead firms in the wind energy sector – while at the same time being evidence of the attractiveness of the Indian market.

Solar cell manufacturing is at an early stage, despite the preference of the government for this sector. Given that solar cell manufacturing in India is more fragmented than the wind power industry, it hardly benefits from economies of scale and remains focussed on the domestic market. It also suffers from Chinese competition. Efforts to boost the sector by local content requirements have led to disputes within the World Trade Organisation (Ernst & Young 2013; NITI Aayog 2015a). A cooperative approach, initiated in 2015, appears more promising: four memoranda of understanding between the government and foreign solar cell producers have been signed so as to facilitate the setting up of branch plants in India. Related to decentralised energy supply, Imergy Power Systems and its US partner Sun Edison have announced the purchase of a large fleet of batteries for application in India. This may prove critical to rural electrification. Some speculate that India may invest up to USD 12 billion in energy storage by 2020 (Buckley 2015).

CLIMATE CHANGE

Like China, India used to be one of the most obdurate opponents to an effective global climate regime – as it would impose key responsibilities for early mitigation on the emerging economies. The government untiringly repeated the mantra of common but differentiated responsibilities and the necessary transfer of financial resources and technologies from the Global North. This line came under increasing attack – especially at the UN Climate Conference in Copenhagen – not only from the West but also from poorer developing countries and island states, as well as from environmental and social movements in India itself. As a result the Indian government offered voluntary mitigation commitments in Copenhagen, much to the chagrin of domestic leftist and ultra-nationalist forces (Betz 2012). Put forward specifically was a reduction of the energy intensity of growth by 20 to 25 per cent by 2020 as compared to the 2005 level.

At the UN Climate Summit in Cancún in 2010, India even appeared prepared to commit itself to binding emissions reductions in what its delegation called an "appropriate legal form" (quoted in Pande 2011: 14), as well as some kind of non-intrusive and non-punitive verification thereof. Yet, coming under pressure from domestic opposition parties, the minister in charge, Jairam Ramesh, had to clarify that he had only "added nuance" to the previously known position of the government. India would not accept internationally binding emission cuts. This minister was soon replaced by a more conservative successor, who fell back fully on the government's former position (*Guardian*, 10 December 2010). Before the UN Climate Summit in Paris, the minister for the environment, Prakash Javadekar, stressed the priority of economic growth and poverty reduction as well as the responsibility of the Global North to act first on climate change mitigation. Practically at the exact same time, he announced also the doubling of coal production up until 2019 (*New York Times*, 26 September 2014).

India is opposed to an ex-ante review of its INDC because it would restrain the country's leeway for future energy consumption and GDP growth. In its INDC, India promises to reduce the emission intensity of GDP by 33 to 35 per cent by 2030 (as compared to the 2005 level), to increase the share of non-fossil electricity generation to 40 per cent of installed capacity and to create a carbon sink of 2.5 to 3 billion tonnes of carbon dioxide equivalent by forest cover. The promised expansion of renewables translates into respective generation capacities of 100 GW for solar energy, 60 GW for wind power and about 15 GW for bioenergy and small hydropower.

The problem with these objectives (and earlier ones announced in Copenhagen) is that meeting them does not require any further efforts being made above and beyond those already planned and implemented. If the planned capacity addition in renewable energies alone comes to fruition, India's emission intensity in 2030 will be less than what the government promised in Paris. Despite the aforementioned challenges in expanding renewable energies, an expert group convened by the Policy Commission did recently calculate the share of renewables in electricity generation to be significantly above 40 per cent by 2030 however (Frank 2016). At the same time, there is no indication of a peak in emissions. Even under the current commitments, India can augment its carbon dioxide emissions indefinitely. This shows that the government intends to retain as much leeway as possible for economic and energy policies that reflect growth objectives, most likely to be based on coal as the main energy source for those.

Climate Action Tracker nevertheless rated India's commitment to climate change mitigation as "medium" in 2015. It however then criticised the fact that the coverage of the commitment was not specified. Most importantly, it did not mention the agricultural sector. Targets were below those to be achieved by envisaged policies, and not consistent with keeping global warming below 2°C.

The current assessment is more positive, suggesting that India's commitment is a fair contribution to meeting the 2°C goal because the emerging economy is by now – and reflecting the aforementioned Draft National Energy Policy – expected to overachieve its objectives regarding the expansion of renewable energies and reduction of emissions intensity. In particular, the 40 per cent target for non-fossil electricity generation is likely to be achieved already in 2020, a decade earlier than promised in Paris. Mention is also made of the intention to disallow diesel- and petrol-fired cars.[1]

NOTE

1. This information is available online at: https://climateactiontracker.org/countries/india.

Chapter 3

Indonesia

Coal is the dominant source of energy in Indonesia, both at present and in the country's build-up programme that is meant to enable it to become one of the top-10 economies in the world. Coal mining has been increasing and will continue on this path in the coming decades, leading to catastrophically high carbon dioxide emissions. We also explain how the country's archipelago landscape configuration complicates both power generation and distribution. The chapter addresses as well the political challenges – an abundance of government bodies with overlapping responsibilities, decentralisation to insufficiently competent authorities and also resource nationalism – that not only hamper the oil and gas upstream sector, but also block the transition to renewable energies.

In particular the tremendous potential for geothermal power is thus foregone, and it appears unlikely that ambitious objectives regarding an energy transition – as formulated in the National Energy Policy – will be met. Efforts to reform the energy sector – especially a much-needed reduction of subsidies – have been only partially successful at best. The chapter also summarises how Indonesia does not possess significant energy-related know-how. With regard to climate change mitigation, it is shown that the country's current commitment is highly insufficient because of the expected large-scale expansion of the coal sector and also ongoing deforestation.

MATERIAL CONDITIONS

Indonesia is relatively well endowed with energy resources, and highly dependent on fossil fuels. Total energy generation, including exports, reached 440 million tonnes of oil equivalent in 2012. Coal accounted for 58 per cent, followed by natural gas (15 per cent), biofuels and waste (13 per cent) and oil (10 per cent). Coal mining increased more than fourfold from 58 million tonnes of oil equivalent in 2002 to 256 million tonnes thereof in 2012. Growth in natural gas production was practically non-existent: up from 65 million tonnes of oil equivalent in 2002 to 67 million tonnes thereof in 2012. Crude oil output fell by 30 per cent. Indonesia also produces geothermal, hydro, solar and wind energy, which account for less than 5 per cent of its total energy generation (IEA 2015b).

In 2012 energy imports totalled almost 53 million tonnes of oil equivalent, consisting of crude oil (38 per cent) and oil products (62 per cent). These imports had risen by 51 per cent during the previous decade, filling the gap between increasing demand and decreasing domestic production. Energy exports amounted to 279 million tonnes of oil equivalent in 2012; mostly coal (81 per cent), but also natural gas (12 per cent) and crude oil (6 per cent) (IEA 2015b). In 2014, Indonesia was the world's largest exporter of coal – measured by weight – and the fifth-largest exporter of LNG (EIA 2015e).

Probably the most characteristic physio-geographical feature of Indonesia is its archipelago landscape configuration. The country encompasses more than 17,000 islands, causing severe challenges for energy supply. Indonesia does not have one national electricity grid, but eight subnational and about 600 local ones. The islands of Bali, Java and Madura, which together use about 80 per cent of the available electricity, form the largest subnational grid – with a capacity of 32 GW. The larger power stations – no matter what source of energy they use – are located almost exclusively on Java. Sumatra's

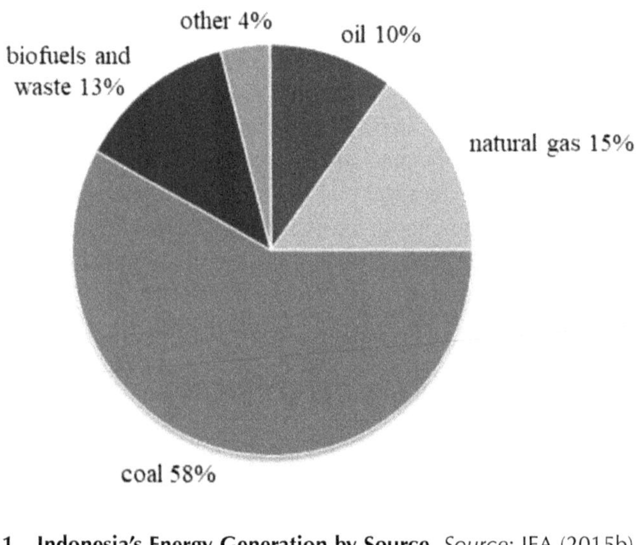

Figure 3.1 Indonesia's Energy Generation by Source. *Source*: IEA (2015b).

grid is the second-largest, reaching a capacity of 6 GW. The third-largest grid is on Kalimantan with 1.8 GW, most of which comes from diesel generators (World Bank 2009).

This situation is rendered even more complicated because Indonesia's hydrocarbon resources are concentrated on the islands of Kalimantan and Sumatra. Java accounts for a much smaller share hereof. The geothermal potential is greatest on Java and Sumatra. Sulawesi and Nusa Tenggara play a secondary role. With regard to hydropower, Papua dominates with a potential of almost 25 GW, followed by Kalimantan (6 GW), Sulawesi (5.5 GW) and Sumatra (4.5 GW) (World Bank 2009). This potential is mostly untapped. Only on Bali and Java are about 2.5 GW – out of a potential 4.5 GW – currently being used (Asian Development Bank 2016).

Indonesia possesses almost 5 billion tonnes of coal, equalling 0.5 per cent of the world's total reserves. Its production amounts to 2.7 per cent of the world's as a whole. The proximity of Indonesia's coal reserves to the coast and relatively good transport infrastructure allow for these tremendous export volumes, which go primarily to China and India. Yet, 60 per cent of Indonesia's coal is of low quality, meaning that its heating power is low.[1] Most of the remaining resources are of medium quality. As we explain below, Indonesia's energy build-up programme relies heavily on coal. Considering the quality of the country's coal reserves, experts argue that the envisaged power expansion programme "will entail environmental calamity" (Bravo and Tsafos 2015: 35). Already today, the coal-fired power stations on Java reach a joint output of almost 18 GW. The World Bank (2009) expects the share of coal in Indonesia's electricity-related carbon dioxide emissions to increase to 72 per cent by 2030.

What is more, the role of coal in Indonesia's energy mix is atypical insofar as coal played a marginal role for electricity generation up until the mid-1980s. Oil and, since the mid-1970s, natural gas used to be the principal energy resources. The rapid rise of coal began in the early years of the new century, and is due primarily to its low price. Oil and gas, conversely, have become more expensive because of depleting (or untapped) domestic resources. The Indonesian landscape, including its territorial waters, contains 60 sedimentary basins. Thirty-six of them are in the west of the country, and well explored. Fourteen are producing oil and gas. Seventy-five per cent of the exploration and production occurs in Aceh (Sumatra), home to Indonesia's oldest large oil fields, and in east Kalimantan (EIA 2015e).

There are eight oil refineries in the country, reaching a joint capacity of slightly more than 1 million barrels a day. The five largest are operated by the state-owned company Pertamina. Chevron – the most important foreign company in Indonesia's oil and gas sector – is developing several deepwater fields in the Kutai Basin in east Kalimantan. They will supply the Bontang

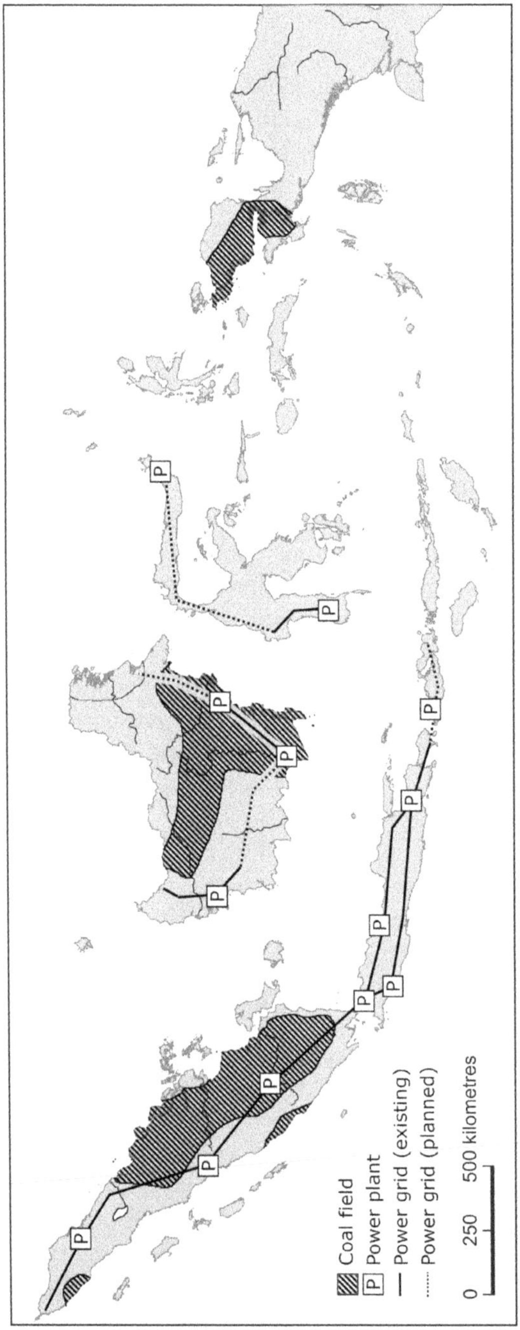

Figure 3.2 Coal Resources and Power Stations in Indonesia. *Source:* Authors' own draft.

LNG plant, which was built in the late 1970s and has since expanded to become one of the largest liquefaction plants in the world with a capacity of 22 million tonnes a year.

Yet, since the early years of this century, no major oil findings have been made in west Indonesia. Production continues to decline. Imports are expected to account for 90 per cent of the country's oil demand by 2030, unless new large oil fields are found in the meantime (Das and Ahlgren 2010). Infrastructural deficiencies, partly due to the archipelago landscape configuration, reinforce this problem: Indonesia has no international oil pipelines and few domestic ones, making maritime trade essential. East Indonesia, meanwhile, remains underexplored, but having considerable potential for future exploration (PwC 2014). The West Natuna Basin in the South China Sea, Bintuni Bay (located in West Papua) and the province of Sulawesi Tengah are emerging as important new offshore exploration areas. However, they are distant from consumers and exploration is – for geological reasons – more costly than it used to be in west Indonesia. The world's major oil companies have, therefore, remained reluctant to invest in Indonesia, even during the height of the global oil price boom about 10 years ago.

The lack of exploration has not only had a negative impact on oil production. Natural gas extraction increased at an annual rate of about 2.4 per cent between the years 2000 and 2010. It peaked with the commissioning of the Tangguh LNG Project. Between 2011 and 2013, production fell by an annual average of 4.4 per cent. One reason for this is that yet-to-be-explored deepwater fields – the Bonaparte Basin in the Arafura Sea and the aforementioned Kutai Basin – throw up considerable geological challenges (EIA 2015e). Natural gas is already being extracted in Kalimantan, Papua and Sulawesi, but these sources cannot supply Java because of a deficit of pipelines and regasification terminals.

Since there is practically no infrastructure to distribute natural gas outside of Java and north Sumatra, the form of it associated with oil production is often flared. Indonesia ranks 12th in global flaring. Experts from the consulting firm McKinsey (2014a) have calculated that resulting losses amount to about USD 100 million a year. In spite of all these challenges, gas-fired power stations on Java reach a joint maximum output of almost 8.5 GW and natural gas is crucial for the expansion of Indonesia's electricity-generation capacities.

There is currently no unconventional hydrocarbon extraction in Indonesia, but the government does appear to be interested in exploring the country's shale oil and gas potential. Indonesia possesses 16.3 trillion cbm of shale gas reserves. The EIA (2015e) estimates that 1.3 trillion cbm of these are recoverable. A major challenge to the growth of the shale industry is the cost of exploration, expected to be four times that in the United States because

Figure 3.3 Oil and Gas Resources in Indonesia. *Source:* Authors' own draft.

Indonesia's deposits are located far deeper below the surface. Furthermore it is estimated that Indonesia possesses coalbed methane reserves of 12.8 trillion cbm, potentially one of the largest worldwide but of rather modest quality. In 2008, the government started awarding blocks for according exploration in east Kalimantan and on Sumatra. So far, production has remained far below initial targets.

A feature that distinguishes Indonesia's energy sector from that of most of the other emerging economies is its endowment with geothermal resources, which could, arguably, generate up to 29 GW of electricity – equalling 40 per cent of the world's estimated potential (EIA 2015e). The Pacific Ring of Fire, a geologically highly active zone where numerous tectonic plates meet, runs through Indonesia's major islands, making the country home to a remarkable concentration of geothermal energy. This energy can be used for electricity production by channelling superheated underground reservoirs to power steam turbines. Indonesia exploits a mere 5 per cent of its geothermal resources (EIA 2015e). Plans to increase the use of renewable energy to at least 23 per cent of the total energy portfolio by 2025 depend on boosting geothermal power. To promote such development, a fast-track electrification plan from 2006 calls for an additional 5 GW of geothermal capacity by 2022, to be operated primarily by private companies.

With regard to other non-carbon based sources of energy, only hydropower plays a somewhat significant role. As noted, the potential for hydropower is greatest in Papua. Presently, total conventional hydropower capacity is merely 4,755 MW. Two relatively large power stations – Cirata and Saguling – are located on Java, reaching maximum outputs of 1,000 and 700 MW respectively. Most other hydropower stations are on Sumatra. Their individual maximum output is less than 350 MW. The Lake Toba Pumped Storage Power Station on Sumatra can generate up to 4,900 MW. Four other pumped storage schemes on Java reach individual capacities of 800 to 1,000 MW.

The potential of biomass, solar and wind energy is considerable, but untapped. Despite high insolation, solar energy is almost exclusively applied for decentralised rural electrification, where it competes with diesel (used for power generators) and kerosene (used for lighting). The present contribution of solar energy to overall electricity generation – about 27 MW – is negligible. At the time of writing, Indonesia's wind park project pipeline totalled 280 MW – thus also amounting to but a small share of the country's according potential. The National Energy Council – a body established in 2007, and usually referred to by its acronym DEN (Dewan Energi Nasional) – expects that biomass holds a potential of up to 55 GW (Ibrahim 2014). Yet, present use of biomass is traditional and hence problematic from an environmental perspective. An estimated 55 per cent of Indonesia's population – almost 130 million people – primarily rely on traditional biomass – mainly wood – for cooking (REN21 2013). Indonesia is also the largest exporter and producer of palm oil in the world, accounting for half of all global production. Six million hectares are being used for palm oil production, with another 4 million hectares to be added.

As this all indicates, Indonesia's electricity generation is only slightly less dependent on hydrocarbons than its energy production is. In 2013, the emerging economy had an estimated 51 GW of installed capacity for electricity generation. Figure 3.4 below shows that 88 per cent of this comes from fossil fuel sources, with the rest being hydroelectric (10 per cent) and geothermal (2 per cent). Other renewables contribute less than 1 per cent. Coal accounts for slightly less than half of the electricity generated from fossil fuels. Natural gas is the second-most important power source. Oil-fired generation has declined along with oil production, but still provides the third-largest share herein (EIA 2015e).

ENERGY DEMAND AND EXPANSION PLANS

Indonesia, with a population of 250 million people, is the most populous country in South East Asia and the fourth-most populous one in the world, behind China, India and the United States. Since the beginning of this century, Indonesia has achieved annual GDP growth rates of about 5 to 6 per cent. As the country industrialises, it uses more fossil fuels – both for electricity generation and for transport. Total primary energy consumption grew by 43 per cent between 2003 and 2013. Domestic demand for natural

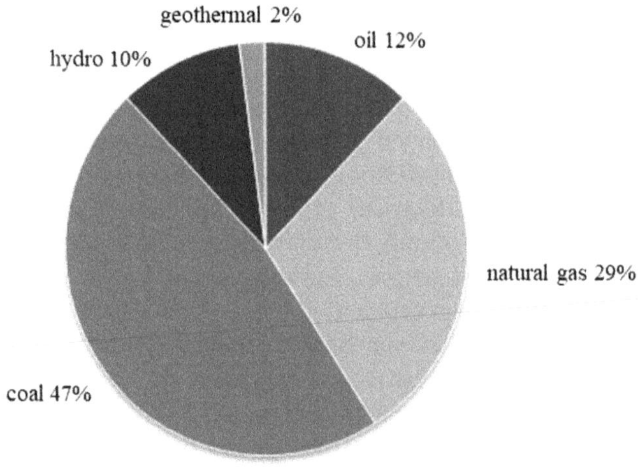

Figure 3.4 Indonesia's Electricity Generation by Source. *Source*: EIA (2015e).

gas has doubled since 2005. That for petrol has increased by 79 per cent (EIA 2015e). Electricity peak demand has approached available capacity, and reserve margins are now inadequate. Blackouts and load shedding occur, particularly beyond Bali and Java. Energy demand is likely to increase even further, as the government aims to transform Indonesia into one of the top-10 economies globally – with per capita income being four times higher by 2025. The National Energy Council argues that per capita electricity consumption will increase from 650 kWh in 2010 to about 7,000 kWh in 2050. The council expects that Indonesia will by then need an electricity generation capacity of 430 to 550 GW, meaning 10 times its current one (Ibrahim 2014).

Considering the rapidly increasing energy demand, Indonesia embarked, as noted, on a fast-track electrification plan in 2006. The first phase of this was about adding 10 GW of coal-based generation by 2010. In spite of considerable delays, 8.1 GW have now been added. The second phase of nearly 18 GW of capacity additions include cleaner sources such as natural gas, geothermal, hydropower and other renewables, but coal-fired power stations still account for more than 60 per cent herein. According to the State Electricity Company (Perusahaan Listrik Negara [PLN], this additional capacity is planned to come online by 2022. Yet there have been considerable delays, as the *Jakarta Post* reported on 13 February 2013 and again on 21 October 2014.

The expansion of Indonesia's energy sector is guided by a presidential decree on the National Energy Policy (Kebijakan Energi Nasional) from 2006, which emphasises diversification, sustainability and the maximisation of domestic resources (President of the Republic 2006a). The National Energy Policy was revised in 2014, setting an ambitious target energy mix of coal (30 per cent), oil (25 per cent), hydropower and renewable energy (23 per cent) and natural gas (22 per cent) for 2025 (President of the Republic 2014). The specificities of the expansion of electricity generation are laid out in the Electricity Power Supply Business Plan (Rencana Usaha Penyediaan Tenaga Listrik [RUPTL]), last updated in 2016. Contradicting the National Energy Policy, it indicates that by 2025 coal will account for half of all the electricity generated in Indonesia. At least 80.5 GW of new capacities have to come online by that year, with 18.2 GW to be constructed by the PLN and 45.7 GW by independent power producers. The remaining 16.6 GW have not been allocated, yet (Ministry of Energy and Mineral Resources 2016).

An earlier version of RUPTL specifies that – in addition to the aforementioned fast-track electrification plan – coal-fired power plants with a capacity of 9,975 MW are to be built in a first phase of expansion. Just 5,700 MW have been delivered so far, however. New capacities in geothermal power (4,000 MW) and hydropower (1,750 MW) and another 3,000 MW of coal-fired electricity are planned in a second phase. The present version of RUPTL calls for 35 GW to be added by 2019. Fifty-six per cent of these additions will be based on coal. Natural gas plays the second-most important role, with a share of 36 per cent. Hydropower (4 per cent) and geothermal energy (2 per cent) remain marginal, contradicting the ambitious objectives set by the National Energy Policy. This planning is largely based on cost calculations: gas-based electricity is twice as expensive as its coal-based variety in Indonesia, while electricity from diesel or fuel oil is four times as expensive (Bravo and Tsafos 2015).

Key to these expansion plans are high-voltage transmission lines. On Sumatra, several power stations are to be built or already under construction on-site at coal mines, most importantly the 2.4 GW Arahan Plant and the 2.5 GW Central Bangko Power Station. The island is to be connected to Java by a high-voltage transmission line by 2019. Bali and Java are to be better connected by that year too, allowing for the decommissioning of some of the diesel generators presently in use so as to cope with the rapidly growing demand on Bali. In total, the government plans to construct almost 50,000 kilometres of new transmission lines by 2020, spending some USD 15 billion for this purpose (Asian Development Bank 2016).

Solar and wind energy do not play a major role for the ongoing expansion of generation capacities. Solar energy is to replace diesel generators in isolated areas, generating 620 MW by 2020. Wind parks are to reach a total of 970 MW by 2025. In other words, the role of these two environmentally friendly sources of electricity will remain negligible. Waste-to-energy plants are being developed on Bali, Bandung, Batam and Jakarta. Considering the waste problems that large Indonesian cities face, waste-to-energy has multiple benefits. However, such plants do not generate electricity at competitive costs because prices remain low – at about USD 0.06 per kWh on Java. The break-even point lies at USD 0.14 to 0.16 per kWh, as the *Jakarta Post* reported on 31 May 2016. The individual capacity of waste-to-energy power stations is also small, reaching 1 to 150 MW only.

First steps to boost bioenergy have been taken. A mandatory biodiesel blending rate of 10 per cent was imposed in 2013. It was raised to 15 per cent in 2015, and to 20 per cent in 2016. Biodiesel and bioethanol, mostly made up of palm oil, are subsidised. The National Biofuel Development Committee has suggested that the government make it mandatory for biofuels to constitute 2 to 2.5 per cent of national fuel consumption (World Bank 2009).

As mentioned in the introductory chapter, the Russian firm Rosatom signed a memorandum of understanding with Badan Tenaga Nuklir Nasional (BATAN) in 2015 and was involved in the construction of an experimental nuclear reactor in Indonesia. The Ministry of Energy and Mineral Resources has issued a white paper on building nuclear power stations whose joint capacity is to reach 5 GW, as the World Nuclear Association (2017a) reports. Bangka Island, located east of Sumatra, is currently the preferred site for the first nuclear power station, which may eventually become operational in the late 2020s. Alternative locations on Java have also been discussed (Ibrahim 2014).

Seeing the long-term goals set by the National Energy Policy, some argue that if the ambitious objectives on renewables are indeed met then Indonesia's use of geothermal energy instead of fossil fuels will equate to the largest carbon dioxide emissions reduction project in the world (Sukarna 2012). Yet Winters and Cawvey, quoting a member of the National Energy Council, describe the objectives on geothermal power as "unrealistic" (2015: 28). By mid-2013, some 30 geothermal projects were not moving ahead. Winters and Cawvey further argue that the government is torn between providing more electricity quickly on the one hand and boosting renewables on the other. Given that the former objective matters more, it is likely that Indonesia will revert to building even more coal-fired power plants than originally planned.

Scenarios developed by Das and Ahlgren (2010) are alarming: the share of coal in power generation may increase to about 70 per cent by 2030. Only if clean-energy technologies are adopted do these contingencies change: biomass and natural gas will partly replace coal for heating purposes. Even in these scenarios, more coal will be used in electricity generation as subcritical coal plants are replaced by new ones based on circulating fluidised bed combustion.[2] Natural gas-based electricity generation is predicted to increase tremendously up until 2020. Afterwards it will decrease, as new technologies – biomass, solar and wind energy – become more widely applied.

Indonesia is also involved in regional infrastructure projects. In order to interlink their natural gas reserves, Indonesia, Malaysia, Myanmar, Singapore, Thailand and Vietnam have together launched the TAGP – which is to consist of 4,500 kilometres of pipeline, mainly situated undersea and worth USD 7 billion. Eight bilateral pipeline projects with a total length of approximately 2,300 kilometres are now operating. Instead of building a network from scratch, the TAGP consists of numerous individual projects that connect existing systems. Regasification terminals are to be built in the context of this joint venture. The TAGP involves numerous hurdles and risks, however. Sovacool (2009) points out that the project will make the South East Asian countries dependent on imported natural gas, because the region only possesses about 8 per cent of global reserves. Apart from that, it appears that the states that participate in the TAGP have not thought about how to deal with potential disputes over transit fees or single countries using their transit position as political leverage. Technical failures and terrorist attacks would be devastating for a centralised distribution system such as the TAGP. The environmental hazards of the TAGP – for example, air emissions from pipelines and soil erosion – are considerable, too.

The TAGP is accompanied by the ASEAN Power Grid (APG), which is to connect the national grids of Cambodia, Indonesia, Laos, Malaysia, the Philippines, Thailand and Vietnam. The APG is – like the TAGP – part of the *ASEAN Vision 2020*, but less progress has been made than initially expected. A first planning document on cross-border connections for the community, finalised in 2003, concludes that a unified electricity grid is not economically viable. It therefore proposes instead 11 bilateral connections. The second planning document, published seven years later, comes to a different conclusion. It calls for more connections being built, and suggests that a single grid be pursued after subregional ones have first been established. After 2020, interconnections for up to 20 GW are to be constructed.

Ahmed et al. (2017) assess the prospects of the APG. The present capacity of cross-border transmissions remains limited to 3.5 GW, though. Indonesia's electricity grids are not interconnected across borders. Although there are considerable geographical obstacles to the APG, its slow progress can mainly be ascribed to incompatible and barely liberalised national electricity markets – as well as to state-owned electricity providers being unwilling or unable to construct the proposed cross-border transmission lines, as Andrews-Speed (2016) explains.

POLITICO-ECONOMIC CONSTELLATIONS

Energy consumption in Indonesia – both electricity and transport fuels – is heavily influenced by subsidies. These were introduced in the 1970s and 1980s as a form of social appeasement and as a way to increase the competitiveness of Indonesian firms. The subsidies were historically financed by the country's high oil-export revenues, which, as mentioned, have declined tremendously in recent decades. They first became untenable in 1997 during the Asian Financial Crisis, when Indonesia's GDP contracted by 13.5 per cent and the government had to devaluate the national currency – which increased the oil-import bill. Subsidy cuts, being part of a structural adjustment programme designed by the International Monetary Fund, meant that the domestic price of diesel rose by 60 per cent, kerosene by 25 per cent and petrol by 71 per cent. The cuts were one among a complex range of factors that triggered protests, which devolved into general rioting – leading to serious violence, and then eventually President Muhammad Suharto's resignation from power.

Later efforts to significantly reduce energy subsidies have been only partially successful. Subsidies were so high that diesel and petrol sometimes cost only half the respective market prices. Since the beginning of this century, there have been five ad hoc price increases, three ad hoc price decreases and two periods during which prices were subject to frequent formula-determined adjustments. Due to subsidies being cut, the price of petrol increased by 33 per cent in March 2005 and by another 88 per cent only seven months later. Price increases have been accompanied by compensation mechanisms such as loans for small businesses, educational support for the families of lower-ranking civil servants and subsidies for basic foodstuffs. There were also temporary unconditional cash transfers to 15.5 million of Indonesia's poorest households for a period of four months after an energy price increase in 2013. Still, Indonesia spent USD 27 billion on energy subsidies in 2013, amounting to almost 11 per cent of state budget expenditure – or 2.5 per cent of GDP. Between 2009 and 2013, fuel subsidies accounted for more public expenditure than infrastructure and social welfare programmes combined (Beaton, Lontoh and Wai-Poi 2017; IEA 2015b). High oil prices up until 2014 and increasing imports made subsidies particularly costly. The government had to reduce diesel and petrol subsidies by more than 30 per cent in 2014. In 2015, it completely eliminated subsidies for petrol – except those resulting from distribution costs beyond central Indonesia – and introduced fixed ones for diesel, which keep the latter commodity's price USD 0.08 below the market one and hence do not compensate for market fluctuations (EIA 2015e).

As a side note, energy subsidies used to be a source of significant fiscal uncertainty because budgeted subsidy expenditure was contingent on overly optimistic assumptions regarding the global oil price, exchange rates and domestic oil production. At the beginning of this decade, politically complex and costly mid-year budget revisions were therefore a frequent necessity (Lontoh, Clarke and Beaton 2014; Lontoh, Beaton and Clarke 2015). Energy subsidies are, moreover, hardly a suitable means for supporting poor households because they purchase only small quantities of electricity and liquid fuels, thus benefitting rather indirectly from the subsidies' impact on the price of staple goods. The richest decile of Indonesian households, meanwhile, took home more than 40 per cent of the fuel subsidies in the early years of the new century (Agustina et al. 2008).

The remaining diesel subsidies have severe negative effects. Although the PLN must not purchase diesel at subsidised prices, there are indications that its regional offices do so. They use this subsidised diesel for electricity generation (Braithwaite et al. 2012), complicating the transition to solar energy in rural areas. Winters and Cawvey (2015) point out that businesspeople who dominate the black market for subsidised diesel – and who are often well connected to politicians – not only oppose a reduction of diesel subsidies. They also work against projects that could create other sources of electricity for rural Indonesia. Structural changes in Indonesia's energy sector would, moreover, conflict with the government's own interests, as the oil and gas industry continues to be a vital contributor to the state budget – accounting for 12 per cent of all state revenues in 2013. About a decade ago, it contributed 21 to 24 per cent thereto (PwC 2014).

The use of electricity is heavily subsidised too – on the basis of distinct customer categories such as businesses, households and industries. Each category is subdivided into small-, medium- and large-scale users. Subsidies to large-scale industrial users were abolished in 2008, but the average electricity tariff has since merely increased by USD 0.05 to 0.06 per kWh. Production costs remain at about USD 0.10 (IEA 2015b). In other words, the government regulates electricity prices below market levels; this forces the PLN to accept losses, which have to then be covered by going into debt, a transfer of state revenues or eventually an increase in prices.

This strategy seemingly reduces the ability of the PLN to invest. The consequence of chronic underinvestment is that energy supply is costly and unreliable. The state-owned electricity provider does not have sufficient means to finance the scheduled expansion of generation and transmission capacities. To ameliorate the financial situation of the PLN, the government decided to raise electricity prices by 15 per cent in 2013. In the following year, medium-scale industrial customers had to pay almost 40 per cent more for electricity; large-scale customers close to 65 per cent. Due to low oil prices, the PLN announced in 2016 a slight reduction in prices. However, its budget gap in 2017 amounted to USD 134.5 billion. It is expected to decline to USD 115.9 billion and USD 42.0 billion in 2018 and 2019, respectively (Asian Development Bank 2016).

Beyond energy subsidies, the Asian Development Bank (2016) criticises the fact that investment in energy generation, transmission and distribution is inadequate and often inefficient, too. The reasons for this are, according to the bank, a lack of financing in conjunction with constraints in policy, planning and implementation arrangements, mostly due to the poor coordination across different government levels. A report by the IEA (2015b) also points out that there are too many institutions with overlapping roles. The coexistence of councils, ministries and task forces complicates policy formulation and implementation. The following government bodies are all involved in the formulation and execution of Indonesia's energy policy:

- The aforementioned National Energy Council is the most important institution. It formulates the national energy policy, and monitors its implementation. The council is chaired by the president, the vice president and the minister of energy and mineral resources. It is composed of members coming from seven ministries. A further eight members are drawn from academia, environmental and consumer groups, and from the industrial sector.
- The Ministry of Energy and Mineral Resources oversees policy formulation and implementation, manages state-owned energy and mining assets,

and evaluates the performance of the sector. It sets the country's thermal coal reference price, and the maximum price for subsidised renewables.

- The Ministry of Finance is engaged in the management of energy subsidies and energy taxation. It must agree on the maximum prices for renewables.
- Environmental standards for resource extraction are set by the Ministry of the Environment and, in case of forests being affected, the Ministry of Forestry.
- The Ministry of Transport is in charge of the transport infrastructure for energy.
- The Ministry of Industry and the Ministry of National Development Planning formulate Indonesia's economic policy, which is closely intertwined with energy policy.

In addition to these high-level institutions, there are numerous subordinate bodies – many of which fail in terms of achieving the original purpose that they were established for and only survive because of baseless announcements of imminent success, as Tanter's (2015) study on BATAN exemplifies. At least some obstacles have been overcome lately, though. For example, the Ministry of Energy and Mineral Resources and the Ministry of Forestry have signed a memorandum of understanding, facilitating geothermal exploration in forests. The Ministry of Energy and Mineral Resources has improved the quality of its data for oil and gas blocks, as well as the transparency of the tender process.

State-owned companies also play a major role in the energy sector, albeit a declining one. The PLN used to be the sole distributor of electricity. This changed in 2016, when the government released the PLN from its obligation to provide distribution and transmission networks in the provinces of Maluku, North Maluku, Papua, West Papua, East Nusa Tenggara and West Nusa Tenggara. Private companies may now build these networks. The PLN has signed contracts with independent power producers and will increasingly do so in the coming years, but it still remains the most important producer of electricity in Indonesia.

Pertamina, the state-owned oil and gas company, operates the vast majority of petrol stations in the country because private companies could not for a long time compete with its subsidised fuel. Pertamina is Indonesia's second-largest oil producer after Chevron, with a share of about 18 per cent of total oil production. Chevron accounts for 47 per cent thereof. Pertamina also accounts for 16 per cent of Indonesia's total natural gas production. Private investors from abroad produce the rest, with BP, ConocoPhillips and Total being the largest (PwC 2014). Pertamina is engaged in LNG and refining activities as well, operating almost all of the country's respective capacities. It has recently expanded into geothermal exploration and exploitation.

The strong role of state-owned enterprises in the energy sector does not make foreign investment impossible, but there is a public discourse according to which Indonesians do not benefit sufficiently from "their" resources. Related to this resource nationalism, the Constitutional Court overthrew a law in 2002 that would have contributed to liberalising the electricity market. Two years later, its judges accused the liberalisation policies pursued by the government of being part of a neoliberal agenda that runs contrary to the values of the constitution. Market pricing would, according to these judges, prevent the government from ensuring energy supply and make Indonesia vulnerable to price manipulation by foreign powers (Mochtar 2015). In 2012, the court argued in a ruling – one that led to the disbandment of the upstream oil and gas regulator – that all natural resources should be "controlled by the state to be exploited to the greatest benefit of the people" (quoted in Stuart 2015: 19).

What is more, the government has imposed a domestic market obligation on large coal producers in response to the increasing domestic demand. It resets caps for the amount of coal that must be allocated for domestic sales each year. This cap stood at 86 million tonnes – out of a production total of 376 million – in 2013. The 2001 Oil and Gas Law includes a domestic market obligation for production-sharing contracts, so that 25 per cent of the natural gas derived from new projects be consumed in Indonesia. Recent renegotiations with Total on the Mahakam Block concluded with a reduction of that company's shares from 50 to 30 per cent. Once production begins, ownership and operatorship will be transferred to Pertamina. Foreign companies that invest in Indonesia also have to cope with time-consuming negotiations over production-sharing contracts, as exemplified by Chevron's recent deepwater and LNG projects (EIA 2015e).

The political context is rendered even more complicated by the fact that local and provincial governments can develop their own plans and regulations on energy. This includes the granting of licences for coal mining and renewable energy projects as well as the supervision of enterprises, including the offices of the PLN. Since 2009, local governments may provide electricity directly to local customers. The reason for this is that following the downfall of the Suharto regime in 1998, the Indonesian legislature passed decentralisation laws. Today, the state is marked by a tug-and-pull dynamic between subnational government units that value discretion and independence and a central government which sees itself as the ultimate guarantor of citizen access to public goods (Turner et al. 2003). Decentralisation has, moreover, led to contradicting and overlapping legislation. Bravo and Tsafos (2015) argue, for instance, that Indonesia's considerably large coalbed methane deposits are barely exploited not only because of problems of land acquisition and insufficient fiscal incentives but also due to inconsistent legal stipulations.

How decentralisation conspires against renewables becomes particularly clear with regard to geothermal power. Since local governments are the owners of all geothermal resources and in charge of related tender processes, the dynamics of the sector depend entirely on them. Yet, local governments lack the technical knowledge necessary to provide useful information to potential investors or to assess the feasibility of energy-generation projects themselves. Tender documents often fail to generate confidence among potential investors – a problem whose relevance can hardly be overestimated, because exploration is the most expensive part of geothermal projects (Winters and Cawley 2015).

In order to ease these problems, the Ministry of Finance has created a Geothermal Fund Facility that supports local governments in preliminary exploration and prepares the tenders in collaboration with them. The fund also subsidises exploration. Yet, as Winters and Cawley (2015) point out, the fund is not allowed to make a loss on investments and thus has been reluctant to support projects for this reason. If exploration fails, the respective local government will be left to carry the losses itself. Local governments, hence, rather use their money for projects that are less risky and more likely to generate political rewards in the short run, such as education and health spending. Civil society action has also been directed against geothermal power because of its potentially hazardous side effects.[3] Given that local governments are particularly vulnerable to such protests, they have become even more reluctant to engage.

Further to that, it appears that firms or individuals participate in tender processes without having sincere interest in exploiting geothermal resources. Their objective is to win tenders and later sell exploration rights to companies that are truly interested in the respective projects. This practice is possible because tenders are awarded to those offering the lowest electricity prices, but the ultimate power-purchasing agreements are not part of the original tender ones – leaving room for later changes. Referring to an unnamed consultant, Winters and Cawvey (2015) suggest that these forms of rent-seeking behaviour make power plants two or three times more expensive than they should be.

ENERGY-RELATED KNOW-HOW

Indonesia hardly develops energy-related technologies, although there are – as indicated with regard to geothermal power – enormous opportunities available for innovative forms of electricity generation. Chances not only for environmentally friendly electricity generation but also innovation-based economic development are largely foregone. CCS has been considered as part of the national energy policy, and some pilot projects do exist. No concrete steps towards its wider implementation have been taken, though. The envisaged expansion of geothermal power depends on foreign investors, whose willingness to do business in Indonesia is limited because of the aforementioned obstacles. In order to advance nuclear power, Russian support is needed.

In 2011, the government announced the Master Plan for the Acceleration and Expansion of Indonesia's Economic Development 2011–2025 (Masterplan Percepatan dan Perluasan Pembangunan Ekonomi Indonesia 2011–2025, MP3EI). The MP3EI rests on three pillars: establishing six economic corridors based on the comparative advantages of the different regions of Indonesia; promoting connectivity within Indonesia and the Association of South East Asian Nations (ASEAN); and, improving human resources and science and technology. According to the MP3EI, almost USD 470 billion will be invested over a period of 14 years, including in energy-related projects. The MP3EI does not, however, provide details on whether – and, if so, how – these investments will be linked to innovation. It mentions a regional innovation cluster focussed on energy in east Kalimantan, but other than that it merely lists infrastructure projects that are to make the country's energy resources accessible. The IEA (2015b) points out that the grand vision of the MP3EI has not been matched by actual implementation.

The transport sector also offers opportunities for climate change mitigation through new technologies. Because of Indonesia's archipelago structure, the domestic shipping fleet consists of around 2,000 vessels. They consume nearly 7,000 barrels of bunker fuel a day. With rapid technological advances in small LNG, this emerging economy could convert large sections of its domestic shipping fleet to LNG. Fuelling just 10 per cent of the shipping fleet with LNG could result in an annual saving of USD 100 million. First steps in this direction would include the setting up of fuelling facilities at major ports, economic incentives to shipping operators so that they make investments to help switch fuels and also supporting regulations for related industries, as experts from McKinsey (2014a) argue.

CLIMATE POLICY

Indonesia's per capita carbon dioxide emissions are much lower than those of the Global North, which has been pointed out by the government frequently. Yet, these emissions are more than twice as high as the average of lower middle-income countries – the group of countries at the same level of economic development as Indonesia. Regardless of questions of responsibility, climate change is particularly problematic for Indonesia because some parts of the country will be more strongly affected than the world as a whole on average. This is not limited to flooding in Jakarta, to which we referred in the introductory chapter. Temperatures will in the course of this century rise by 2.0 to 2.5°C on average and by 2.5 to 3.0°C on Borneo and Sumatra, according to a high-growth scenario by the IPCC.[4] Fish stocks are expected to deplete, or move away from Indonesian territorial waters. Numerous small islands will be at increased risk due to rising sea levels. The risk of severe flooding will be reinforced by El Niño and La Niña occurring more often. Both phenomena account for wave height variations of 2 to 5 metres.

During the last 10 years, the government has published numerous papers on climate change and climate change mitigation, most importantly the Climate Change Sectoral Roadmap (Republic of Indonesia 2009), the Framework for Nationally Appropriate Mitigation Actions (Ministry of National Development Planning 2013) and the National Action Plan Addressing Climate Change (Republic of Indonesia 2007). They refer to deforestation – together with emissions from peatland, the main source of carbon dioxide emissions in Indonesia – and to the energy sector. The respective papers list issues to be addressed, but they are thin in terms of implementation strategies and targets whose achievement could be concretely evaluated. The following measures have been considered, and their implementation could at least partly be assessed:

- Energy efficiency, including fuel efficiency standards, is to be enhanced. Government planning calls for energy savings of 17 per cent by 2025. From then on, energy consumption is to grow more slowly than the GDP. In 2015, the government launched a national campaign to reduce energy use by 10 per cent by 2019.
- Renewable energy development will be fostered in accordance with the (partly contradictory) electricity build-up plans summarised above.
- The use of biofuels is to be increased. For this purpose, the aforementioned blending rates and subsidies were introduced.
- Transport demand management and the quality of public transport are to be improved. Bus rapid transit systems will be introduced in 23 cities. Rail-based mass transit will be developed in nine metropolitan areas.

Indonesia's current energy policy undermines any responsible climate one. The expansion of coal-fired power generation is not accompanied by clean-coal technology and CCS. Subsidies still lead to an inefficient use of energy. Natural gas – often believed to make a vital contribution to climate change mitigation, insofar as it replaces coal – will only be less harmful than coal if certain risks are avoided. In addition to bringing toxic rock fragments and contaminated water to the surface, the extraction of natural gas can come along with methane emissions, too. Methane is 23 times more harmful as a greenhouse gas than carbon dioxide. Natural gas processing usually causes emissions of carbon dioxide and nitrogen oxides, which tend to be excluded from related assessments of emissions.

At the UN Climate Summit in Copenhagen, Indonesia pledged to reduce its greenhouse gas emissions by 26 per cent until 2020, as compared to a business-as-usual scenario; or, by 41 per cent if international support is provided. According to the INDC, Indonesia will reduce its greenhouse gas emissions by 29 per cent until 2030, also as compared to a business-as-usual scenario. Provided that international support is granted, the reduction will amount to 41 per cent. This is to be achieved through adaptation in land use – mainly a slowdown of deforestation – and energy generation, most importantly the (unrealistic) goal of a 23 per cent share of renewable energies by 2025. The INDC remains vague on how Indonesia plans to carry out the necessary adaptation.

Against this backdrop, it should not be surprising that Climate Action Tracker rates Indonesia's commitment to climate change mitigation as "highly insufficient". It is inconsistent with limiting warming to less than 2°C, unless other countries make comparably greater efforts. The reason for this assessment is that Indonesia's emissions from energy generation are set to increase rapidly up until the early 2030s. They have risen more rapidly in the second half of this decade than in the first. To be consistent with the Paris Agreement, they ought to stabilise by 2030 – if not begin to decline. As noted, electricity-related carbon dioxide emissions are certain to increase tremendously, whereas the expansion of renewables has been much lower than planned.[5]

Climate Action Tracker now points out that there has been a peak in forestry-related emissions, whereas it earlier suggested that Indonesia is the only main deforesting country where a strong increase in deforestation emissions can be expected. Three weeks of intense deforestation induced by fires in this emerging economy can generate as many greenhouse gas emissions as the entire German economy in the course of a year. Trends indicate a loss of 25 per cent of the current forest area by 2030 under present policy settings. A government agency charged with restoring peatlands, where half the country's deforestation happens, has been created. It struggles to implement meaningful policies, also because local and regional politicians support palm oil and pulp-and-paper companies, arguing that they create jobs (and sometimes being bribed by these companies, too). The central government lacks the means to control the entire territory so as to guarantee that large companies do indeed respect the limits on deforestation that it has imposed. For the same reason, it appears impossible to put a halt to deforestation by small-scale farmers. Corresponding support by Western states has fallen behind initial promises, at the *Economist* reported on 26 November 2016.

At least from a short-term perspective, deforestation is indeed in Indonesia's economic interest; as are the expansion of coal-fired power generation and the export of coal. Even immediately occurring negative environmental side effects are difficult to measure and often almost impossible to attribute to a single source. The people who suffer from these side effects are not politically influential, but they are in need of cheap electricity and jobs in coal mines or on palm oil plantations. This dilemma is exemplified by Indonesia's moratorium on the conversion of natural forests to land used for palm oil production. As said, Indonesia is the largest exporter and producer of palm oil in the world; a scenario-based analysis by Yusuf, Roos and Horridge (2018) concludes that the moratorium will reduce economic growth in the coming decades, in particular in those parts of the country that heavily depend on palm oil being produced. International transfers – payments made to avoid deforestation at about USD 10 per tonne of unreleased carbon dioxide emissions – could more than compensate for these welfare losses. Whether international donors will be willing to pay and how it can be guaranteed that the money will be used to the benefit of the Indonesian people are, of course, much more difficult questions to answer.

NOTES

1. Coal is divided into two categories: hard/black and soft/brown. These categories are subdivided into anthracite, bituminous, sub-bituminous and lignite coal, which reflect the respective age – and thus the degree of carbonisation. Coal is considered of low quality when it has a high content of ash, moisture or sulphur, reducing heating power and increasing toxic emissions.
2. The circulating fluidised bed is a developing technology for coal combustion designed to achieve fewer emissions. By using this technology, arguably up to 95 per cent of pollutants can be absorbed – especially by sulphur-absorbing chemicals such as dolomite or limestone.
3. Geothermal sites may contain poisonous gases that can escape through holes drilled for exploration. Water pumped from underground reservoirs for use at geothermal power stations is hot and often contains high levels of mercury, sulphur and other hazardous substances. Underground water may become contaminated. The necessary cooling systems, moreover, consume vast quantities of water.
4. This and other scenarios are available online at: www.ipcc-data.org.
5. This information is available online at: http://climateactiontracker.org/countries/indonesia.html.

Chapter 4

South Korea

South Korea's energy sector is dominated by fossil fuels and nuclear power. Both depend on imported resources. The potential for renewables used to be seen as low. Their current role remains marginal. Considering the dominance of heavy industries, several South Korean governments have advanced plans to decrease energy intensity and substitute coal and oil by natural gas, nuclear power and renewable energies. We show that corresponding targets have not been met, in spite of suitable policy instruments being developed.

Afterwards, we explain that large, state-owned energy companies and the slow liberalisation of energy markets work against the necessary transition to renewables. Private firms, in fear of their competitiveness, are largely opposed to the government's green growth paradigm. They have also slowed down climate-friendly initiatives such as carbon dioxide taxation. We then elaborate on South Korea's sophisticated strategy of world leadership in green technologies, which aims at the domestic market and large-scale exports. Against this backdrop, we argue that efforts to mitigate climate change are economically motivated. It appears unlikely that South Korea will meet its INDC, and, even if it does, the country's contribution to climate change mitigation will remain highly insufficient.

MATERIAL CONDITIONS

In contrast to other Organisation for Economic Co-operation and Development (OECD) countries, South Korea's energy consumption is still increasing (IEA 2012). Its energy consumption and emission profile are disproportionate to the level of economic development: South Korea has a small population of only 50 million people, but the ninth-largest national energy consumption in the world. Its per capita consumption is higher than in any other non-Western, non-oil-exporting country, except for Singapore (Shift Project n.d.). The industrial sector accounts for 33 per cent of GDP, and heavy industries are of outstanding relevance (production of cement, steel and petrochemicals, as well as shipbuilding). This and the high share of manufactured products explain the still-significant rise of energy consumption and greenhouse gas emissions: by 2.6 and 2.2 per cent a year, respectively (Enerdata 2017). Primary energy consumption in South Korea is dominated by oil (44 per cent) and coal (29 per cent). Natural gas (14 per cent) and nuclear energy (11 per cent) also play considerable roles. The share of renewable energy is negligible (2 per cent), and consists overwhelmingly of biofuels and waste.[1]

Due to its own very limited domestic resources, South Korea is one of the top energy importers in the world – in particular, for coal, oil and gas. Energy imports account for 92 per cent of consumption (IEA 2012). In 2015 South Korea's crude oil imports amounted to 2.8 million barrel a day, the fifth-largest volume in the world. Main suppliers were Saudi Arabia, Kuwait, Iraq, Qatar and the United Arab Emirates, in that order (EIA 2017). Efforts to diversify suppliers have been barely successful. To protect against interruptions of supply and oil price fluctuations as well as merely to follow international regulations, private and state-owned oil companies have built up massive stockpiles of 370 million barrels (IEA 2014c). Domestic production accounts for less than 1 per cent of consumption, but the emerging economy

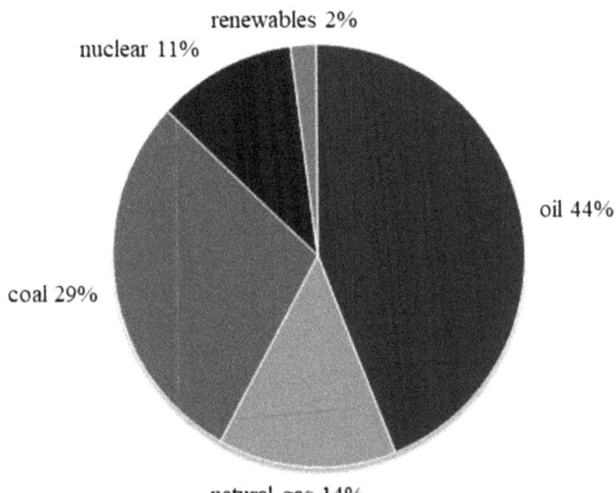

Figure 4.1 South Korea's Primary Energy Consumption by Source. *Source*: IEA data, available online at: www.iea.org/statistics.

hosts 3 of the 10-largest oil refineries in the world, namely GS Caltex, the S-Oil Corporation and SK Energy – with respective capacities of 669,000, 785,000 and 840,000 barrels a day. These meet the domestic demand, and enable South Korea to export large and indeed increasing quantities of refined products – in particular diesel and jet fuel, mostly to Australia, China, Japan, Singapore and Taiwan (EIA 2017).

Because of the increasing relevance of power stations fired by natural gas, South Korea has become the world's second-largest consumer of LNG, behind only Japan. Main import sources are Qatar, Indonesia, Oman, Malaysia and Nigeria, in that order. In South Korea, there are five regasification facilities currently in operation. Four of them are owned by the Korea Gas Corporation (KOGAS), which maintains a near monopoly on the purchase, import and distribution of natural gas. The fifth facility is run by KOGAS in tandem with a private partner. Most of the LNG imports are burnt in one of eight power stations, which reach a capacity of more than 1 GW each. Domestic reserves, meanwhile, are minimal, amounting to 5.7 billion cbm. Domestic production is low: it reached 198 million cbm in 2015, and the country's only active natural gas field will stop producing in 2019 because of depleting resources. It remains to be seen whether offshore exploration, which first started in 2012, will lead to commercially viable findings (EIA 2017). Dilapidated nuclear reactors returning to service and the cheaper price of coal imports explain why South Korea's natural gas imports declined from their apex of 47.7 million tonnes of oil equivalent in 2013 to 39.3 million tonnes thereof in 2015, according to data from the IEA.

Coal reserves are equally negligible, amounting to only 326 million tonnes (BP 2017) – less than annual consumption. Hence consumption is met almost entirely by imports, which have by now reached 135 million tonnes – making South Korea the fourth-largest coal importer in the world (EIA 2017). The government has successively closed down domestic mines ever since the late 1980s, as their productivity was in rapid decline. Imports come mainly from Australia, Indonesia and Russia. Sixty per cent of them are used for electricity generation. There are six giant coal-fired power stations in South Korea: Boryeong, Dangjin, Hadong, Samcheonpo, Taean and Yeongheung, all of which have a capacity of 3 GW or more. Their first units were built from 1983 onwards. The other units are of recent design, and among the most efficient plants in the world – meaning ultra-supercritical (Ecofys 2016).

As a reaction to the growing dependence on energy imports, South Korea launched an aggressive strategy of developing and expanding nuclear energy in the late 1970s – made possible by its strong industrial base and competence in services. The country has to import all of the uranium required to feed its power plants. A 30-year cooperation agreement with the United States, extended by another 20 years in 2015, has proven critical in this regard. Under this agreement, this emerging economy is not allowed to enrich uranium or to recycle used nuclear fuel, which would diminish dependence on imported uranium but also bring the country closer to nuclear weapons capacity (World Nuclear Association 2017b). Currently there are 25 reactors in operation, including two very large ones: Hanbit (6.1 GW) and Hanul (6.2 GW). Altogether, South Korea's nuclear power stations reach a maximum output of 23 GW (EIA 2017).

The potential of renewable energy was for a long time regarded as modest. About 70 per cent of South Korean territory is mountainous. Installing solar and wind power facilities there would be costly. In the flat parts of the country, population density is very high, leaving little space for power plants to be built there. Offshore wind power is hampered because of concerns over fishing and military installations (Duffield 2014). Yet, recent assessments reach more optimistic conclusions. According to the Renewable Energy White Paper from 2014, the technical power generation potential from renewable energy amounts to more than 12,000 TWh a year – which is 22 times the country's current electricity use. As noted, the actual share of renewables in electricity generation is, however, low. While waste dominates the applied renewable energy capacity with a share of about two-thirds, modern forms of energy generation play a much less important role: biomass reaches a share of 13 per cent, followed by hydropower (10 per cent), solar (5 per cent) and wind energy (3 per cent) (In-Ha 2014). Total installed wind power is just slightly more than 1 GW. Six medium-sized hydropower stations reach a joint maximum output of only 3.9 GW.

ENERGY DEMAND AND EXPANSION PLANS

The industrial sector dominates energy demand in South Korea, being responsible for 27 per cent of total consumption, followed by the residential sector (26 per cent), transport (20 per cent) and the services sector (12 per cent). In the past two decades, the share of industry herein has significantly increased (Chung 2014). In 2008, a major change in South Korea's economic, energy and environmental policies occurred, when President Lee Myung-bak proclaimed a vision of "low carbon, green growth". The rationale behind this vision is that South Korea's development trajectory is not sustainable, and ought to shift to one of quality-oriented growth – with a focus on energy independence and sustainability achieved through the increasing use of renewable energies (Choi 2015). To realise this vision the government seeks to play an active role herein, in line with its one during the economic take-off from the 1960s to 1980s.

The National Strategy for Green Growth (2009–2050) along with consecutive five-year plans for green growth as well as the Framework Act on Low Carbon, Green Growth, passed by the National Assembly in 2010, are the policy tools to achieve this economic transition. Main targets are to build a low-carbon society (by supporting renewable energies, eco-labelling and standards on energy efficiency), to enhance energy security – which means energy independence by 2050 – and to foster supposedly green technologies such as advanced light-water reactors, CCS and electric cars. Twenty-five per cent of expenditure for research and development is to be diverted to green sectors – an objective that has almost been achieved (if a very broad definition of green sectors is applied), as we summarise below. Green finance and the creation of a carbon dioxide market are the other pillars of the shift towards low carbon, green growth.

Regardless of the envisaged shift, the first national Energy Master Plan – which covered the years from 2008 to 2030 – projected an increase of primary energy demand by nearly 50 per cent, equalling an annual growth rate of 3.7 per cent and a total consumption of 342.8 million tonnes of carbon dioxide equivalent by the latter year. Energy intensity was expected to fall by 2.1 per cent annually, in a business-as-usual scenario. Under a more aggressive one, energy demand would only increase by 1.1 per cent a year and energy intensity decline by an annual 2.6 per cent. The plan also stated that the share of fossil fuels would decline to 75 per cent by 2030, or to 61 per cent in the more aggressive scenario. Only natural gas was projected to gain in importance – but not coal and oil – and the share of renewables should reach 11 per cent by 2030.

The Energy Master Plan aimed at turning the renewable energies industry into a growth engine for the entire South Korean economy. It proposed market-based but government-supported incentives, in order to foster capacities in research and development related to renewable energies – that to make South Korean industries more competitive, also with regard to potential exports. From 2002 to 2012, the government supported renewable energies by differentiated feed-in tariffs determined according to location, installed capacity and power source. Power producers received subsidies based on a complicated formula for a period of 15 to 20 years in a descending manner. Prices were lowest for small hydro, followed by for bio and ocean energy. Prices for solar power were considerably higher (Global Green Growth Institute 2015). Feed-in tariffs have allowed for a rapid development of renewable capacity, albeit at a very low level.

These came at a heavy and growing cost though, especially in the case of solar power. In 2012, the government thus discontinued the feed-in tariffs and switched to a renewable portfolio standard (RPS). Under this scheme power producers are obliged to attain a share of 10 per cent of renewables in their respective generation by 2022, starting with 2 per cent in 2012 and rising by 0.5 per cent each year thereafter. Power producers are free to choose their sources (but required to give a certain preference to solar energy), and able to buy and sell their obligations in case of under- or overachievement. Still, targets for renewables have not been met: their share in 2006 (2.2 per cent) fell short of what should have been achieved by then (3 per cent). This gap became larger at the beginning of the present decade (2.75 versus 5 per cent). Until today, unmet obligations for renewables only continue to grow (Global Green Growth Institute 2015; Yoon and Sim 2015) – making it unlikely that the rather modest target set for 2030 will ultimately be met. The Energy Master Plan also envisaged an enormous expansion of nuclear power, to 41 per cent of energy supply by 2035. This target has since been reduced to 29 per cent (Ministry of Trade, Industry and Energy 2014); the reason for this being the Fukushima incident, followed by safety concerns regarding South Korean reactors, too (Duffield 2014).

The Energy Master Plan faced considerable challenges in its implementation. The planned introduction of near market prices was not achieved due to political reluctance over tariff hikes (more on this later), a factor that partly explains why energy consumption grew more intensely than planned (by 3.3 instead of 2.6 per cent a year from 2007 to 2012) – especially in the industrial sector. The target for energy-intensity reduction was, therefore, missed by a rather wide margin of 13 per cent (Ministry of Trade, Industry and Energy 2014). Peak demand for electricity also increased so much that power outages became more frequent, most prominently in September 2011 – when an outage affected more than 2 million homes (Chung 2014).

The second national Energy Master Plan, which replaced its predecessor in 2014, did not constitute a change of direction. It was based on an expected annual GDP growth rate of 2.8 per cent between 2011 and 2035 and a reduction of the energy intensity of growth by 30 per cent, to be achieved mainly by a slowdown in the growth of industrial demand. Overall energy demand would increase from 275.7 million tonnes of carbon dioxide equivalent in 2011 to 377.9 million tonnes thereof in 2035. Renewable energies were to grow fastest (4.4 per cent a year), followed by nuclear power (3.3 per cent) and natural gas (1.9 per cent). Oil demand, meanwhile, should decline by an annual rate of 0.15 per cent. This would have roughly translated into a stagnation of the absolute oil demand. All other sources of energy were to grow in absolute terms, with the clear majority of energy generation remaining being based on fossil fuels. Renewable energies, including hydropower, would not renegotiate their marginal role. How the targets for individual energy sources should be met was not discussed in depth in the plan (Ministry of Trade, Industry and Energy 2014).

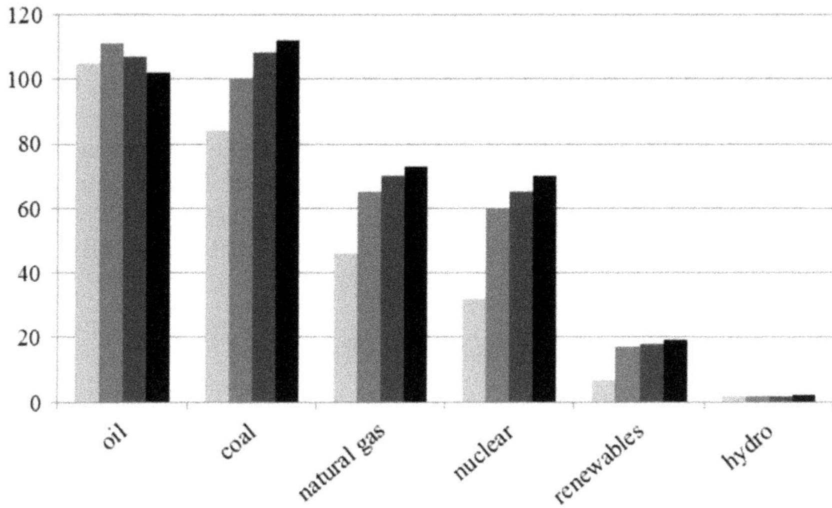

Figure 4.2 South Korea's Present and Future Energy Demand by Source, in million tonnes of oil equivalent. *Source*: Ministry of Trade, Industry and Energy (2014). *Note*: The years of reference are 2011, 2025, 2030 and 2035 (from left to right).

Recently, a major overhaul of energy planning took place with Moon Jae-in becoming South Korea's new president. The 8th Basic Plan for Electricity Supply and Demand foresees a lowering of peak energy demand in 2030 from 113 GW to 102 GW, a considerable increase in the share of renewables to 20 per cent of electricity supply by that year, as well as a reduction of nuclear power to 24 per cent and of coal to 36 per cent. The revised projections are in line with lower GDP growth and the promises of the newly elected president to move away from nuclear energy. As a consequence of this policy shift, the country's oldest nuclear reactor was shut down in 2017. Eleven other ageing reactors will be decommissioned in due course. President Moon also promised to close 10 coal-fired power plants (equal to 11 per cent of coal-based power generation), as the newspaper *Hankyoreh* reported on 16 July 2017. This decision reflects public concern over air pollution (Cornot-Gandophe 2018). However, coal-fired power generation continues to expand – instead of being replaced by natural gas (Paik 2018).

Beyond the national scale, and to make up for the dearth of domestic oil and gas reserves, South Korean private and state-owned companies have engaged in exploration projects overseas. The Korea National Oil Corporation alone holds assets in 29 blocks all over the world. The largest investments have been made in Iraq (USD 47 billion), Canada (USD 4 billion), Uzbekistan (USD 3 billion) and the UK (USD 3 billion). These acquisitions have been financed by a heavy credit intake. State-owned KOGAS has invested in exploration and production abroad. It has also acquired shares in foreign companies to boost LNG supplies. KOGAS sought to secure a quarter of South Korea's domestic natural gas demand by imports, but the government decided already in 2013 to divest some of the currently running projects because of cost overruns and mounting debts. The same applies to oil assets abroad too: little of the oil extracted in the just-mentioned countries is shipped to South Korea, while the debts of the Korea National Oil Corporation and other involved companies have shot up (EIA 2017; Jeon 2011).

POLITICO-ECONOMIC CONSTELLATIONS

The institutional set-up for energy policy is more centralised in South Korea than it is in the other emerging economies. The Ministry of Trade, Industry and Energy is the sole body responsible for energy policy, including climate change mitigation, energy saving, price control, resource development and the general supervision of the industrial sector. The Korea Electricity Commission has the task of regulating the electric power sector, but it operates under the Ministry of Trade, Industry and Energy. There is also an affiliated think tank, the Korea Energy Economics Institute, which recommends policies to the ministry. When the first Energy Master Plan was drafted, the government formed the National Energy Committee as a deliberative and decision-making organ – being composed of about 60 experts from government bodies, the private sector and non-governmental organisations (NGOs). In order to advance the new policy strategy, the Presidential Committee on Green Growth was created. It develops and implements green growth efforts that involve various ministries and public agencies. The committee includes private sector representatives, which is characteristic of the traditional pattern of policymaking in South Korea – where the government closely cooperates with the private sector in deciding on the country's economic course.

Whereas the political centralisation that results from these procedures allows for a high degree of efficiency, especially when it comes to adjusting energy policy to short-term changes in resource prices, South Korea's energy sector suffers from the dominance of large corporations, price subsidies and related problems. Measures to deregulate the energy sector have been taken. Starting in 1997, petroleum prices were liberalised. They are now fully determined by market principles. Tax rates for diesel and petrol have been adjusted to economic cycles, meaning that they have been lowered during periods of downturn. Domestically mined coal continues to be subsidised; but, in accordance with its green growth strategy, the government intends to phase out this funding by 2020 (IEA 2012). The natural gas market, meanwhile, has not been liberalised. The resulting lack of competition accounts for end-user prices above the OECD average. KOGAS – a de facto state-owned and state-operated company – is the sole importer, with the exception of direct ones by some large industrial consumers.

Hence, KOGAS is responsible for 95 per cent of South Korea's natural gas imports. With regard to distribution, around 30 private companies have been granted exclusive retail rights for clearly delineated regions. All distributors are supplied by KOGAS at prices that it sets itself. The government of President Park Geun-hye announced the deregulation of the natural gas sector by 2025, allowing private companies to import and sell LNG. In addition to cheaper prices, the objective of this liberalisation is to increase the overall share of natural gas, which is critical to cutting back on coal. As noted, President Moon announced major shifts in energy policy – but, up to now, there have been no indications that these will include a more rapid liberalisation of the natural gas market.

Electricity generation, transmission and distribution are controlled by the state-owned Korea Electric Power Corporation (KEPCO). The generation wing of KEPCO was split into six subsidiaries in 2001 in order to enhance efficiency, by promoting competition and responsible management. Public divestment of these subsidiaries has been postponed due to a lack of interest among potential private investors. KEPCO still functions as the sole distributor of electricity. Retail prices for it are heavily subsidised to protect poor customers and, even more so, agriculture and the industrial sector, for which prices are set at 33 to 50 per cent of the average one – thus being the lowest among all OECD countries (Chung 2014; IEA 2012). In 2016, the government proposed a partial privatisation of KEPCO and announced that it would allow more private participation in power generation and distribution, the time frame of which has yet to be decided. As in the case of the natural gas market, the government expects liberalisation to bring down prices for end users, who pay considerably more for electricity than citizens in other OECD countries do (*Chosun Ilbo*, 15 January 2016) – unless they are beneficiaries of the just-mentioned subsidies.

In addition to the dominance of a handful of corporations, which, as seen, are often state-owned, there is a strong nexus between public agencies and the energy sector; this becomes most apparent concerning the nuclear industry. In government publications nuclear energy is praised as a clean and reliable source of energy, which may explain the relative neglect of renewables. Unlike nuclear power, renewable energies do not have an influential lobby group in South Korea (Yon and Sim 2015). Even the Ministry of Trade, Industry and Energy (2014) complains that a narrow political focus on nuclear self-sufficiency has resulted in decreased transparency within the industry itself, also leading to insufficient safety procedures.

Besides its political connections, two factors explain the strength of the nuclear sector and the relative neglect of renewable energies. First, South Korea's development path has hitherto been based on capital- and energy-intensive sectors. The country relies on cheap and reliable energy. Decentralised renewable energies have remained a niche in an energy system marked by centralised solutions, as advanced by a small number of large-scale firms. Second, nuclear power promises electricity at a low price. We are aware that debates on electricity prices are always highly politicised, especially with regard to negative externalities. Still, under current conditions, nuclear energy costs USD 0.053 per kWh in South Korea, compared to USD 0.058 for coal, USD 0.086 for solar, USD 0.092 for wind and USD 0.107 for natural gas (*Hankyoreh*, 22 January 2017). Park et al. (2016) argue that the cost gap between nuclear energy and renewables is even larger.

After the Fukushima incident in 2011, nuclear energy suffered from a reputational setback. The incident gave rise to a strong and nationwide anti-nuclear movement. The rejection of nuclear energy was intensified by the detection of falsified quality-control documents for hundreds of components of reactors. However, this scandal only lead to the temporary shutdown of two reactors in 2012. In 2013, and also in reaction to the Fukushima incident, the National Energy Committee recommended reducing reliance on nuclear energy from 29 to 22 per cent of total power generation. The Park government eventually adopted the upper end of this range, meaning that there will be no reduction of planned nuclear capacity addition until 2024 (Duffield 2014). Still, it was argued already in 2013 that a turn away from nuclear energy was likely to cause a shortage of electricity (Reuters, 11 August 2013). As noted, President Moon appears to be seeking a change in this regard. It remains to be seen whether more nuclear power stations will be shut down during his presidency, or whether he will slow down the expansion of the sector at least.

Another challenge for renewable energies is resistance by the domestic industry in general, meaning beyond just the nuclear sector. The reason for this is a somewhat justified concern about hardware being imported from China at the expense of domestic firms (Yoon and Sim 2015). The circumscribed power of proponents of renewable energies, meanwhile, is best exemplified by the considerable gap between the aforementioned targets for renewable deployment and actual achievements. Public expenditure for renewables is much lower than it is for nuclear energy. What is more power generators have mostly failed to meet their RPS targets, and corresponding penalties have so far been too low to induce a change in their business strategies (Global Green Growth Institute 2015). Renewable deployment is, according to Kim and Mathews (2016), also hindered by the near monopoly of KEPCO as a distributor of electricity. Only a liberalisation of the electricity market will end this disadvantaged position of providers of renewable energy. There is, moreover, growing resistance to large wind parks because of noise pollution. Permits to build wind farms are difficult to obtain due to complex regulations, leading to long delays.

Still, there are some support mechanisms in place for renewable energies. As noted, feed-in tariffs were introduced in 2001 (but later cancelled and replaced by the RPS). Subsidies for the installation of solar panels on rooftops of residential buildings have been granted since 2009. Since 2011, it has been mandatory for large public buildings to meet at least 10 per cent of their energy consumption from renewable sources (In-Ha 2014). A Target Management Scheme for greenhouse gas emissions was established in 2011. The scheme consisted of agreements negotiated by the government with the largest emitting companies. It aimed to reduce energy use per company by 12.5 per cent below a business-as-usual scenario. The scheme operated with rather mild sanctions, but it nevertheless achieved satisfactory private participation and promising results (Duffield 2014; Global Green Growth Institute 2015). In light of this success, the government launched the Emissions Trading Scheme in 2015, copying the European model of tradeable emission allowances. A larger role of mass transit, a rising share of biodiesel and fuel efficiency standards applied to all vehicles sold from 2015 onwards are additional components of this energy-efficiency strategy. Incentives have also been provided for the construction of energy-efficient buildings and the introduction of energy-efficient industrial processes (Duffield 2014; Jones and Yoo 2011; Mathews 2012; Republic of Korea 2010).

In view of the rapid and far-reaching transformation intended by the green growth strategy and the economic as well as political weight of traditional energy-intensive industries, it does not come as a surprise that private companies have mobilised resistance to the green growth strategy itself – in particular, against the planned reduction of the energy intensity of production, the cap-and-trade system and the envisaged cost- and market-conforming electricity prices. Private companies refer to a potential loss of competitiveness vis-à-vis foreign enterprises and the lack of specific mitigation capacities in the domestic industrial sector as justification for their position (Chung 2014; Global Green Growth Institute 2015). This resistance has frequently been boosted by international developments unfavourable to climate change mitigation. For example, the reform of the electricity and natural gas markets – promised since 2009 – came to a standstill when the international oil price recovered from its nadir in 2008. Sales taxes on diesel and petrol had to be adjusted downwards. This response to the feared backlash from poor consumers and industry demonstrates that the transition to a less energy-intensive path is difficult even under a centralised, efficient government.

Beyond the fact that there are powerful forces pushing back against a shift towards green growth, it should be noted that the strategy is in fact motivated by economic considerations. This may explain why South Korea's pro-business conservative party has become the main promoter of the green growth strategy. Green growth is meant to overcome the country's dependence on imported energy resources. Energy conservation, improved efficiency and the deployment of renewables will all have the same effect. Environmental standards in major receiving markets for South Korean manufactured goods – especially regarding vehicle fuel – have come along with the threat of penalties for non-compliance. The traditional economic model of capital-, energy- and export-intensive development has lost part of its utility in general. Green growth, conversely, offers the chance to give rise to new growth engines. At the same time, South Korea is quite vulnerable to climate change and this certainly contributes as a secondary motivation to the green growth paradigm. The average temperature rise in its major cities was 1.5°C over the past century, which is twice the global average. Rains have become heavier and more concentrated, resulting in natural disasters. Sea levels have risen, also causing challenges for this emerging economy (Global Green Growth Institute 2015).

ENERGY-RELATED KNOW-HOW

South Korea is among the OECD countries with the highest share of investment in research and development, spending 4.2 per cent of its GDP thereon (compared to an OECD average of 2.6 per cent). As part of its green growth strategy, the emerging economy expanded public investment in research for energy and environment technologies from KRW 2 trillion in 2009 to 3.5 trillion in 2013, which equalled 20 per cent of all public investment in research and development in South Korea in that year. Three-quarters of this money was destined for 10 core green technologies, among them batteries, CCS and electric cars (Paeng 2010). South Korea's potential in terms of energy-related know-how is boosted because most corresponding research and development is carried out by private firms and focussed on applied solutions.

South Korean firms benefit from public funding, but have themselves eagerly taken up new opportunities. This is probably best exemplified by enterprises from the shipbuilding sector, which have successfully expanded into offshore wind power (Park 2016). The photovoltaics industry, conversely, suffers from strong Chinese competition. It appears that it was rather Chinese than domestic enterprises who benefitted from the aforementioned solar feed-in tariff (Global Green Growth Institute 2015; REN21 2016).

In 2009 various institutions for energy research were integrated into a single agency, the Korea Institute of Energy Technology Evaluation and

Planning (KETEP), placed under the authority of the Ministry of Knowledge Economy (which then became the Ministry of Trade, Industry and Energy in 2013). KETEP is in charge of developing a strategic plan for investment in research related to sectors with high market potential. It finalised a corresponding road map in 2011. This specifies 15 green energy sectors (such as biofuels, CCS and smart grids), 88 strategic items (for example, offshore wind power systems and small nuclear reactors) and 288 key technologies (multi-fuel fuel processors and organic semiconductors, among others) for South Korea to focus on. It provides strategic directions (including specifications on capital requirements), a year-by-year schedule for research and development, and a related commercialisation strategy. The objective of the road map and strategic plan is to make South Korea a global leader in green technologies and to capture a significant portion of the market for green technologies (18 per cent of the global green energy market by 2030, per the aforementioned road map).

The road map and strategic plan were followed by a strategy for green technology commercialisation, which narrows the approach somewhat down by focussing on technologies in seven specific sectors: CCS, fuel cells, hybrid cars, LED lighting, light-water reactors, smart-grid systems and solar cells. Another 20 areas of technology have been singled out for medium- and long-term support. In particular, CCS has benefitted from generous public financing since 2009. Pilot projects have been completed. Storage sites – including some offshore – have been explored, and combustion technology has been developed (IEA 2012).

While green technologies are being developed rapidly, expertise in nuclear forms is already spectacular. South Korea's first nuclear reactor became operational in 1978. In the early stages of its nuclear programme, cooperation with foreign companies was essential – but the government initiated a self-reliance programme for plant design and engineering as well as non-critical components in 1985 already. Today, indigenous technologies are used even in the critical components of new reactors. This high level of technological sophistication has allowed the emerging economy to even export nuclear technology: in 2009 a South Korean consortium won a bid to construct reactors in the the United Arab Emirates (UAE), prevailing over French and Japanese competitors (O'Donnell 2013). South Korea is now marketing its reactor types in Africa and Latin America, too. It has, furthermore, signed corresponding cooperation agreements with Brazil, Jordan, Kenya and Ukraine, among other countries. The Ministry of Knowledge Economy/ Ministry of Trade, Industry and Energy has declared that it aims to export 80 nuclear reactors with full intellectual property rights by 2030, making South Korea the third-largest supplier of such technology globally (World Nuclear Association 2017b).

CLIMATE POLICY

As this all shows, South Korea has adapted its energy and economic policies in a way that does indeed contribute to mitigating climate change – although corresponding policy shifts appear to be economically motivated, and not driven by concerns over the environment (at least not as first priority). Nevertheless, some institutional adjustments have been made with explicit regard to climate change mitigation. In 2009, an Energy Efficiency Bureau was established within the Ministry of Knowledge Economy. We mentioned the Target Management Scheme already above. It covered about 600 large-scale industry facilities. More than 90 per cent of the companies participating in the first round hereof achieved their energy consumption targets. Those not meeting the targets were not sanctioned, but rather supported in building up the required technical capacity for future mitigation (Global Green Growth Institute 2015).

Yet, targets set by the government for the reduction of emissions and energy intensity vary considerably from one sector to another, being relatively low for industry in general (reduction by 18 per cent) while high for buildings (27 per cent) and transport (34 per cent). Capital- and energy-intensive industrial sectors have to fulfil the lowest reductions: steel production by 6.5 per cent and shipbuilding by 6.7 per cent. Some of these targets were subsequently adjusted downwards after protest from business associations (Global Green Growth Institute 2015).

National industry has been lobbying against more ambitious targets. During the deliberations for the first five-year plan for green growth, which took place between 2009 and 2011, three targets were proposed: a reduction of emissions by 21, 27 or 30 per cent until 2020. The industry opted for the lowest of the three targets, arguing, as noted above, that a higher one would endanger international competitiveness. Business associations pointed out that South Korea still has a long way to go in its commercialising of climate mitigation technologies, and that the preponderance of energy-intensive sectors speaks against overly ambitious targets. In eventually opting for a more ambitious goal, the government has been supported by civil society organisations. Hence the 30 per cent target has since been adopted and broken down to economic sectors, with due regard for heavy industry (Global Green Growth Institute 2015).

The industrial lobby managed to delay the proposed tax on carbon dioxide emissions for vehicles. Legislation for a general carbon dioxide emissions trading scheme was, however, passed in 2012, leading to the step-by-step introduction of it from 2015 onwards as the aforementioned Emissions Trading Scheme. It replaced the Target Management Scheme, and is expected to address 68 per cent of South Korea's greenhouse gas emissions. Targets for each industry have been set, but the majority of allowances will be allocated free of charge initially. Auctions are planned for a later stage. Sanctions within this new scheme are modest. Small companies are not yet covered, and the industrial sector is favoured in the distribution of certificates (Chung 2014; Kim 2015; Korea Energy Agency 2015).

In international fora, South Korea has made commitments to climate change mitigation. After adopting the green growth strategy, it declared at the UN Climate Conference in Copenhagen its willingness to reduce greenhouse gas emissions by 30 per cent by 2020 as compared to a business-as-usual scenario. This was, at the time of the summit, an ambitious target, comparable to those proposed by the member states of the EU. Still, on the base of the assumed GDP growth rate, it implied only a small reduction in absolute emissions: 545 million tonnes of carbon dioxide equivalent in 2020, compared to 550 million tonnes thereof in 2010. From 2009 until 2012, emissions did indeed rise more slowly than before. Since then, a slow decline in emissions has been achieved – although energy consumption has been rising (Enerdata 2017; Olivier, Schure and Peters 2017). Part of this change can be attributed to the dip in economic growth after 2011. It appears that a definitive decoupling of energy consumption and emissions from economic growth has yet to happen.

In 2015 South Korea communicated its INDC ahead of the UN Climate Conference in Paris, promising a reduction of carbon dioxide emissions by 37 per cent by 2030 (meaning to a total of 536 million tonnes of carbon dioxide equivalent, or 18 per cent less than in 2010). Operating with rather strong assumptions about growth in GDP of 3 per cent a year and a constant share of manufacturing, absolute business-as-usual emissions for 2030 were set at 850 million tonnes of carbon dioxide equivalent – thus generating the impression that South Korea is genuinely willing to make a significant commitment to climate change mitigation. The government also refers to the high share of manufacturing in the country's economy and the public resistance to a larger deployment of nuclear energy as an explanation for why its commitments cannot be more ambitious. As emissions in 2014 were already above the pledged limit for 2030, fulfilling the INDC would necessitate a drastic reduction occurring soon.

Even if South Korea somehow manages to meet the goals set by its INDC, this contribution will still be "highly insufficient" according to Climate Action Tracker. This means that it is not consistent with international fairness in view of the country's development level. Climate Action Tracker initially rated South Korea's contribution as "inadequate", criticising the fact that land-use change was left out of the INDC. Its inclusion would have made the commitment even weaker. Clarity over what business as usual implies as well as the stringency of the emissions trading system have also been called into question, as a large share of reduction commitments – 11 of the aforementioned 37 per cent – is to be achieved through international market mechanisms, whose eventual coming about is highly uncertain. Climate Action Tracker now recognises the efforts undertaken by the Moon administration to expand renewable energies, but criticises the inability of the government to close coal-fired plants planned or under construction because this means that South Korea will remain heavily dependent on fossil fuels. Absolute greenhouse gas emissions by 2030 will be twice as much as in 1990.[2]

NOTES

1. This information is available online at: www.iea.org/statistics.
2. This information is available online at: http://climateactiontracker.org/countries/south-korea.

Chapter 5

Brazil

Oil dominates energy generation in Brazil, whereas the electricity sector is relatively clean, being marked by a 64 per cent share of hydropower. We explain how this configuration results from natural conditions, and elaborate on Brazil's fleet of power stations. Further to that, we point out that there is a tremendous (and untapped) potential for solar and wind energy, which could complement hydropower.

The chapter summarises Brazil's expansion plans for energy generation, which have turned from ambitious to moderate ones due to the country's recent economic crisis. Other topics dealt with are regional cooperation on energy, the changing role of the state-owned oil giant Petróleo Brasileiro – or Petrobras, for short – and its involvement in corruption, as well as Brazil's world leadership in energy-related know-how (biofuels, hydropower and off-shore oil). The chapter summarises the INDC of this emerging economy and its sound national policies on climate change mitigation, many of which were drafted by the Lula da Silva government. We conclude by arguing that even modestly positive developments in this regard have suffered a setback during the Michel Temer presidency. Announcements made by the new president, Jair Bolsonaro, suggest that Brazil will now refrain from making any contribution to climate change mitigation altogether.

MATERIAL CONDITIONS

Oil is the most important energy source for Brazil with a share of 65 per cent in primary energy generation, followed by hydropower (15 per cent), natural gas (9 per cent), fuel ethanol (6 per cent) and non-hydro renewables (5 per cent).[1] However, Brazil's electricity generation is relatively clean, which sets it apart from the other emerging economies. Hydropower accounts for 64 per cent of the electricity generated in Brazil, followed by natural gas (13 per cent), biomass (8 per cent), oil (5 per cent), coal (4 per cent), wind (4 per cent) and nuclear energy (2 per cent) (Ministério de Minas e Energia 2016).

Another particularity of Brazil is that it does not run short of oil and gas. The emerging economy has 15 billion barrels of proven oil reserves. About 95 per cent of Brazil's oil is located offshore, mostly off the coast of the federal states of Rio de Janeiro and São Paulo. Slightly more than 70 per cent of crude oil production takes place there. Proven natural gas reserves amount to 453 billion cbm (EIA 2015f). Yet, the plentiful oil deposits found in 2006 are located in ultra-deepwater areas and below thick salt layers, making their exploitation difficult. The Tupi Field, which is the largest recently found field, is some 5,500 metres below the ocean's surface; the second-largest one, the Jupiter Field, slightly more than 5,000 metres below it. Ultra-deepwater drilling in the Gulf of Mexico, for example, starts at a depth of 1,500 metres. Hence, it is not surprising that the first oil well drilled into the pre-salt layers there took over a year to be completed and cost USD 240 million – and indication of the obstacles that Brazil faces regarding its own reserves.

The semi-statal company Petrobras, which is the global technology leader for pre-salt extraction, nowadays drills a similar well in 60 days, at a cost of USD 60 million. Today, about a quarter of Brazil's oil production comes from these reserves – up from only 0.4 per cent in 2008. Petrobras expects that half of its oil output will come from pre-salt deposits by 2020.

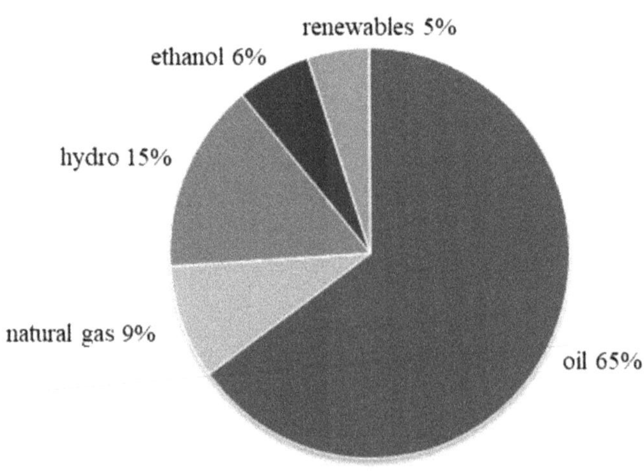

Figure 5.1 Brazil's Primary Energy Generation by Source. *Source*: Shift Project data, available online at: www.tps-data-portal.org/all-datasets.

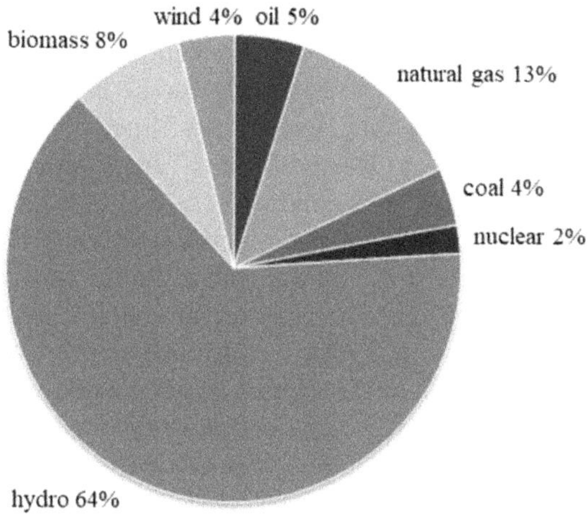

Figure 5.2 Brazil's Electricity Generation by Source. *Source*: Ministério de Minas e Energia (2016).

As a result of increasing production and almost stable domestic consumption, Brazil became a net exporter of oil in 2009. Presently, exports and imports are almost equal (EIA 2015f). Petrobras's investment plans have been slowed down, however, by a severe economic and political crisis, including corruption scandals that involve the company itself (more on this later).

The Campos Basin, off the coast of Rio de Janeiro, is the major deposit site of natural gas. Onshore resources are concentrated in the Amazon area, which hosts about 72 per cent of Brazil's terrestrial natural gas. Considering that 73 per cent of Brazil's natural gas extraction comes from offshore fields, what happens onshore is much less relevant; also because the remote Amazon area is far away from consumers and lacks sufficient infrastructure to provide access to resources found there (EIA 2015f). The IEA (2013) estimates conventional natural gas resources recoverable in Brazil to be 26 times higher than the country's current proven reserves. The emerging economy probably also hosts the world's tenth-largest technically recoverable shale gas resources, at 6.9 trillion cbm. Their exploitation would be relatively expensive, though. Proven tight oil reserves amount to 5.3 billion barrels, less than in Argentina, Colombia and Venezuela (EIA 2013a).

As noted, Brazil's infrastructure for natural gas is not fully integrated across the entire country. The Brazilian pipeline network is almost 12,000 kilometres long. It mostly covers the north-east, south and south-east of the country, as shown by Figure 5.3 further below. For years, different subnational systems were not interconnected at all. Since 2010, the north-eastern and south-eastern markets have been linked by a 1,400-kilometre pipeline, which runs from the city of Bahia to Rio de Janeiro. In 2009, Petrobras completed the Urucu Pipeline, which links Urucu – an isolated location in the Amazon rainforest – to Manaus. The Urucu Pipeline is expected to facilitate the development of local natural gas resources. On the international scale, the Gasbol Pipeline – inaugurated in 1999 – plays a critical role. The pipeline reaches a capacity of 13 billion cbm a year. Sixty-eight per cent of Brazil's natural gas imports were transported through it in 2014 (EIA 2015f). Gasbol connects Santa Cruz in Bolivia to Corumbá in Brazil and continues to São Paulo, where the gas-fired Piratininga Power Station reaches an output of about 4 GW. Most other large gas-fired power stations are located in the federal state of Rio de Janeiro, close to Brazil's own developed natural gas reserves.

Unlike the other emerging economies, Brazil meets a significant share of its energy demand for transport with biofuels. The country is the second-largest producer of ethanol in the world, behind the United States. Today, cars used in Brazil have engines that burn petrol blended with 20 to 25 per cent of ethanol or pure ethanol. The ethanol programme ProAlcool was launched in response to the oil crises of the 1970s. In 1975, the required share of ethanol in petrol was raised to 20 per cent. It had only been 0.3 to 3.5 per cent before then. The government guaranteed high prices and provided cheap credit for producers. After a supply crisis in the mid-1980s, due to high sugar prices and low oil ones, major investments were made to foster the cultivation of sugar cane. At the beginning of this century, most car manufacturers in Brazil began to introduce vehicles capable of running on any combination of ethanol and petrol. Sales of flex-fuel cars surpassed those of ones with conventional engines for the first time in 2004 (Pinguelli Rosa, Pires and Muylaert 2009). Soybeans have become an additional basis for biofuel production.

Yet, biofuels suffer from market fluctuations. Brazil's biofuel imports, coming mostly from the United States, have undergone a considerable shift in terms of both price and quantity in recent years. Droughts have forced the emerging economy to import ethanol. High sugar prices entice producers to switch to sugar production instead of ethanol. Given that the bulk of the costs of ethanol are embedded in those of sugar cane, price fluctuations for the latter have a very strong impact on those of the former. If sugar cane is not processed quickly, the crop soon rots. The seasonality of the sugar cane harvest causes, therefore, a shortage during a number of months each year. In addition to that, non-governmental organisations, various researchers and even the World Bank (2007) warn that the increasing use of biofuels – both as derived from soybeans (used in blended diesel) and from sugar cane (used in blended petrol) – negatively affects food security in developing countries.[2]

We address the relationship between biofuels and climate change in due course, which is not as clear cut as one might think.

Returning to electricity generation, coal plays a minor role for the Brazilian energy sector. The country possesses more coal resources than its neighbours: about 32 billion tonnes, located almost entirely in Rio Grande do Sul and Santa Catarina. These suffer from a high ash and sulphur content, making them barely useful in energy generation. There are only three medium-sized coal-fired power plants in Brazil: the Jorge Lacerda Station (857 MW) in the south, and the Porto do Itaqui Station (365 MW) and the Porto do Pecém Station (1,095 MW) in the north-east. Coal used in iron and steel works is mostly imported from Australia and the United States because production thereof requires its high-quality variant. Another obstacle to using domestic coal is the country's insufficient railway network. Coal is typically transported by rail, but Brazil's railway network is not nationally integrated. Coal-fired power stations are, thus, typically built on the coast and supplied by ship.

Brazil possesses vast uranium resources, which amount to 278,000 tonnes – or about 5 per cent of the world total. Production in domestic uranium fuel fabrication plants is expected to rise to levels sufficient to cover the needs of the country's three-existing power stations (Kidd 2013). The Brazilian nuclear power programme was launched in the early 1970s. The Angra 1 and Angra 2 Power Stations near Rio de Janeiro were built in collaboration with West Germany. Due to macroeconomic and technical problems, Angra 1 commenced its operations only in 1982, reaching a modest maximum output of 640 MW. Angra 2 became operational in the year 2000, with a capacity of 1,350 MW. Angra 3 might go online before the end of the current decade, generating up to 1,405 MW.

Hydropower power stations are, as noted, much more important than those that rely on coal and uranium. They account for 89 GW of Brazil's generation capacity of 134 GW, one of the highest ratios in the world. The world's second-largest hydroelectric plant by generation is the 14 GW Itaipu Station on the Paraná River, which Brazil operates together with Paraguay. A total of 25 hydropower stations distributed across the country reach an individual output of more than 1 GW. This capacity cannot be fully exploited because of a lack of water in times of drought. Many existing and planned hydropower stations are located far away from the main centres of demand, which explains high transmission losses – that among others factors, such as the poor maintenance of the grid. Together with distribution losses, they have remained at 16 to 17 per cent of generated electricity since the beginning of this century (World Bank 2014). Brazilian authorities expect transmission losses to increase, in fact, because of power stations being built in peripheral areas (Ministério de Minas e Energia 2010).

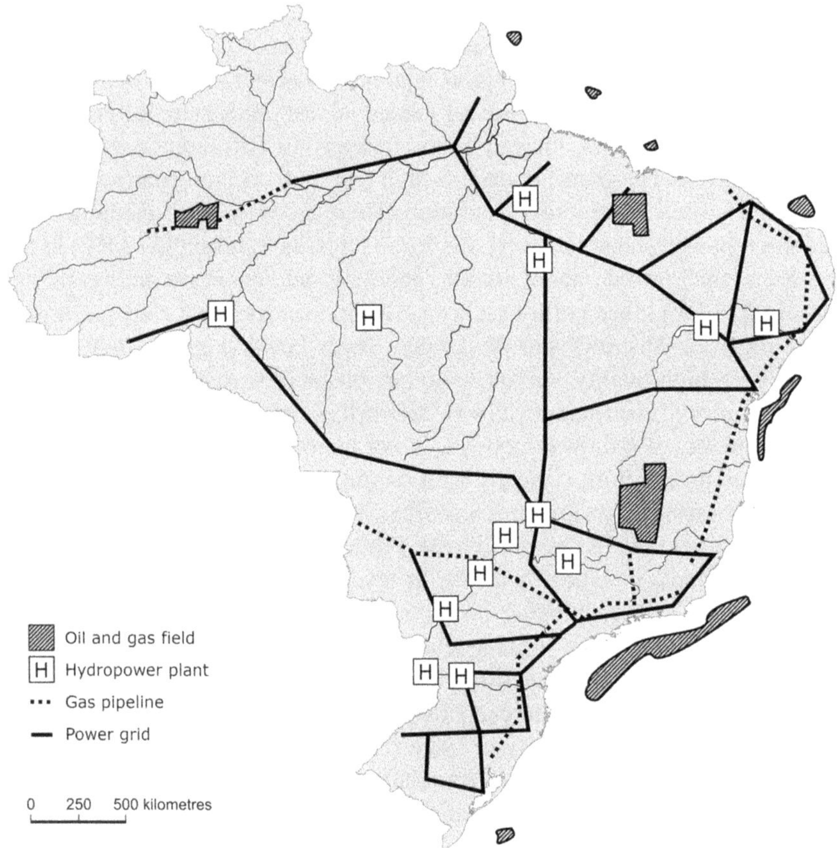

Figure 5.3 Energy-Related Infrastructure and Resources in Brazil. *Source*: Authors' own draft.

The electricity transmission system is currently the source of most power interruptions, along with electricity theft (*O Globo*, 30 November 2017). The expansion of the grid and its maintenance are behind schedule. These delays are problematic because a functioning network is meant to compensate for the regional variation in hydropower generation. A major transmission line from Belo Horizonte to Salvador is currently being constructed. Belém is already linked to Brasília and São Paulo. High voltage lines from the centre of the country to the south and south-east are planned. To the best of our knowledge, no such plans exist for better interlinking Brazil's economic heartland – Rio de Janeiro and São Paulo – with the north and north-east of the country.

Natural conditions in Brazil are somewhat favourable to non-hydro renewables. In particular, the semi-arid north-east is highly suitable for solar energy (Ramos Martins et al. 2008a). As pointed out in the Brazilian Atlas of

Solar Energy, this part of the country benefits from the influence of tropical high-pressure zones almost all year round. They provide stable conditions of low nebulosity and high solar irradiation (Bueno Pereira et al. 2006). Solar energy also has great potential in cities because it can contribute to meeting peak demand there. The commercial areas of tropical and subtropical cities are marked by a mid-day peak of demand due to the intense use of air conditioning. This peak coincides with the maximum solar irradiance. Peak loads are higher in summer for the same reason, also boosting the case for solar energy (Ramos Martins et al. 2008b). Yet, solar energy presently accounts for a negligible share of electricity generation: a meagre 0.01 per cent.

What is more, solar energy can be applied beyond larger urban areas: in the sparsely populated Amazon area, there are hundreds of mini-grids, being operated by independent power producers and local state utilities. Mini-grids extend over some 45 per cent of this vast area, but they supply energy to only 3 per cent of the population – usually relying on diesel generators. Most of the sites where they operate are not easily accessible, increasing costs and decreasing the reliability of supply. If the mini-grids were powered by combined diesel–solar generators, costs could be reduced significantly (Ramos Martins et al. 2008b).

Biomass could also expand considerably. It is, as noted, already today the third-most important source of electricity in Brazil. It is based on by-products from the paper, sugar and wood industries and might further benefit from sugar cane cultivation. If the cane were not burnt before the harvest, leaves and tops could be collected and incinerated to produce steam to generate electricity. Alternatively, they could be converted into alcohol fuel – thus decreasing the severe air pollution caused by burning sugar cane. Brazil's potential for wind energy is 145 GW, according to the Atlas of Brazil's Wind Power Potential (Camargo do Amarante et al. 2001). Unlike hydropower, it is concentrated in the north-east (75 GW), south (23 GW) and south-east (30 GW) – meaning in relative proximity to the vast majority of commercial and private customers. The existing wind parks are located in these parts of the country. Their joint capacity has increased considerably: from 927 MW in 2010 to 8,715 MW in 2015. Wind energy is particularly attractive because the highest potential wind energy generation – at least in the north-east – occurs in spring, when the reservoirs at hydropower stations are typically at their lowest levels.

With regard to regional cooperation, infrastructure plays a limiting role. Pipelines for natural gas connect Brazil only with Argentina (capacity of 4 billion cbm a year) and Bolivia (13 billion cbm a year) (Gomes 2014). The overly ambitious Great Pipeline of the South, as launched by Argentinean president Nestor Kirchner and his Venezuelan counterpart Hugh Chávez, aimed at interlinking Argentina, Brazil and Venezuela. It failed because of the risks associated with the nationalisation of energy resources and critical infrastructure, as well as Brazil's focus on developing its own domestic offshore reserves (*Clarín*, 20 October 2006 and 19 November 2007). The Mercosul Energy Ring – first discussed in 2005, and presented by the *Economist* (18 August 2005) as a major option for the integration of Chile into the subregion – has not become a reality either. It would begin with a 1,200-kilometre pipeline from south Peru to north Chile, linking Peru's Camisea Fields to a network that would also include Argentina, Brazil and Uruguay. Bolivia and Paraguay could connect at a later stage. In the context of IIRSA, less ambitious projects are being advanced: natural gas pipelines in north Argentina, which is already linked to Bolivia, and cross-border ones connecting south Brazil with Bolivia, Paraguay and Uruguay (COSIPLAN 2016). The outcome of these initiatives is uncertain, though.

The prospects for regional cooperation on electricity are slightly better. Brazil's grid is linked with the ones of Argentina, Paraguay, Uruguay and Venezuela. Two large projects of bilateral power generation are framed by IIRSA. The Garabi-Panambi Hydroelectric Complex, which Argentina and Brazil have pursued based on different plans ever since the early 1970s, is now scheduled to reach a maximum output of 2.2 GW – this to be generated at two yet-to-be-built dams. The Madeira River Hydroelectric Complex consists of two dams and two power stations in Brazil (the Jirau Power Station with a capacity of 3,750 MW and the Santo Antonio Power Station with a capacity of 3,580 MW), with a third on the border of Bolivia and Brazil – which is expected to reach about 3 GW – and a fourth on Bolivian territory itself. Whereas Argentina and Brazil would share the electricity generated at Garabi and Panambi equally, most of that generated by the Madeira River Hydroelectric Complex is designated for use in Brazil alone (COSIPLAN 2016).

ENERGY DEMAND AND EXPANSION PLANS

Before the onset of the country's recent economic and political crisis, the IEA (2013) predicted that Brazil would become a net oil and gas exporter by 2035 – mainly because of the exploitation of its pre-salt reserves. Oil production was expected to triple over the next two decades. Together with new refining capacity, this would enable Brazil to meet all its oil product needs. Yet, the emerging economy suffers from the fact that most domestic refineries are built to process light crude oil, whereas the variant extracted from the pre-salt reserves is of heavy grades. It cannot be processed in the country itself. Domestic natural gas production does not meet the present demand, either. Brazil, hence, imports natural gas from Bolivia by pipeline and, in much smaller quantities, via LNG terminals from Nigeria, Qatar, Spain and Trinidad and Tobago.

During the last two decades, Brazil has more than doubled its electricity generation. Growth in demand has been driven by government-supported rural electrification and social tariff programmes. After da Silva of the Workers Party (Partido dos Trabalhadores [PT]) won the presidential elections in 2003, energy accessibility increased significantly. With an electrification rate of presently 99 per cent, Brazil has practically reached universal access, except for some remote rural areas that are only sparsely populated. The growing middle class, moreover, accounts for different patterns of consumption: the use of air conditioning, fridges and washing machines has increased considerably, and private car ownership has been pushing fuels demand (Luomi 2014).

Pre-crisis assessments, accordingly, predicted a considerable expansion of electricity-generation capacities. The IEA (2013) argued that the country would need 6 GW of new capacity each year up until 2035. Following this scenario, total generation capacities would double by 2035. More than half of the expansion was expected to come from hydropower: nearly 70 GW, of which only 7 GW would be small-scale. Natural gas and wind energy were predicted to account for 30 GW of new capacities each. Bioenergy and photovoltaic capacities were also projected to expand, to a total of 16 and 8 GW respectively.

The key planning documents that guide Brazil's energy policy are two long-term ones – the Plano Nacional de Energia 2030 (PNE 2030) and the Plano Nacional de Energia 2050 (PNE 2050) – plus the annually updated medium-term energy expansion plan, or, per its Portuguese acronym, the PDE. PNE 2030, which predates the current economic crisis and is, therefore, based on a high GDP growth rate, projects electricity demand to rise from 321 TWh in 2000 to 847 or even 1,244 TWh by 2030. Total energy production ought to increase by 5 per cent a year from 2005 to 2010, and by an annual 3.4 to 3.6 per cent between 2010 and 2030. The share of oil in Brazil's energy consumption is to drop to 30 per cent by 2030. Natural gas will rise to 15 per cent. Ethanol is expected to increase its share to 18.5 per cent. Electricity generation will see some shifts, with hydropower increasing to a capacity of 156 GW by 2030 and coal (6 GW), natural gas (7 GW) and nuclear power (21 GW) still playing a much less important role. Non-hydro renewable energies will increase to a total capacity of 20 GW (Ministério de Minas e Energia 2007).

At the time of writing, only certain sections of PNE 2050 were publicly available. The increase in electricity demand projected by that plan is even higher: to 1,624 TWh by 2050. Natural gas consumption in the residential sector is expected to increase by a factor of 13. Energy demand by the transport sector will rise by 2 per cent a year (Ministério de Minas e Energia 2014b). According to Brazil's last pre-crisis PDE, oil production is to increase from 2 million barrels a day in 2013 to 4.9 million in 2023. Natural gas output will rise from 77.2 million cbm a day to 148.8 million within the same time frame. Electricity-generation capacities are to rise from 125 to 196 GW. The most recent PDEs predict, meanwhile, a rather moderate increase in energy demand, reflecting the fact that they were written in the midst of the economic crisis. According to these, the shift in energy generation will be more gradual – with the respective shares of hydropower and oil declining and those of ethanol and natural gas increasing.

According to other pre-crisis plans, 33 new hydropower stations should have become operational between 2015 and 2019 – increasing the total hydropower capacity to 117 GW (Ministério de Minas e Energia 2010). Most notable among these projects is the Belo Monte Hydropower Station in Pará, situated in the eastern Amazon Basin. Upon completion, it will reach a maximum output of 11 GW – thus being the third-largest hydroelectric plant in the world, behind China's Three Gorges Dam and the aforementioned Itaipu Dam. Hydropower remains dominant in the national energy mix, but, in recent years, thermal generation has seen a rapid increase, too. Since the beginning of this century, two new coal-fired power stations and 16 new natural gas-fired power stations have gone online. A 1.5 GW plant, fired by natural gas, is to become operational in Sergipe by 2025.

Together with non-hydro renewables, natural gas is being increasingly used to compensate for variations in hydropower generation in times of drought. This strategy dates back to the beginning of this century. In the summer of 2000 and 2001, low rainfall led to a prolonged national energy crisis. Twelve years later, another water shortage occurred and made electricity prices on the free market shoot up to BRL 823 per megawatt-hour (MWh), up from a 12-month low of BRL 129. Still, hydropower is by far the most economically sound – and, arguably, also environmentally friendly – source of electricity for Brazil. Following calculations by Carvalho and Sauer (2009), its cost is about a third of electricity from natural gas and 35 to 40 per cent of that from coal-fired and nuclear power stations. A recent study by Oliveira (2014) concludes that price differences are even more substantive, with large-scale hydropower generating electricity at BRL 85 per MWh and oil-powered thermal energy – which stands at the other extreme – costing BRL 600.

The rapid drop in prices for non-hydro renewables is most apparent for wind power. In open auctions, wind energy now outcompetes electricity generation based on coal and natural gas. Luomi (2014) reports that prices in wind energy-exclusive auctions fell by almost 30 per cent from 2009 until 2013, which is not exceptional in global comparison but nevertheless still impressive. In 2015, there were 262 wind parks operating in the country and 437 either being planned or already under construction. Another outstanding development in the non-hydro renewables sector is South America's largest solar power installation, which will reach a maximum output of 292 MW; it is being constructed by the Italian company Enel in Piauí. Another park, built by Enel in the federal state of Bahia, became operational in early 2017, and generates up to 158 MW.

The decision to complete the Angra 3 Power Station was taken by the PT, which also decided to start the enrichment of uranium on a commercial scale, using technologies developed in Brazil. This was meant to enable the country to acquire the complete nuclear fuel cycle, and to build a multipurpose research reactor. Nuclear energy is presented by the government as a clean source of energy that helps to mitigate climate change. More importantly, nuclear power stations could compensate for hydropower ones in times of drought if built in sufficient quantity (Patti 2015). Rosatom has shown interest in participating in the expansion of Brazil's nuclear sector (*O Globo*, 7 June 2013). Given the lack of available public funds for building Brazilian-owned plants, Rosatom may construct, own and operate nuclear power stations in the country – thereafter selling electricity at a guaranteed price. Nonetheless, nuclear energy does not play a major role in post-crisis expansion plans – at least not when measuring its relevance by contribution in terms of MW.

In comparison to the other emerging economies, the electricity sector in Brazil is fairly liberalised, with it following market mechanisms. All sources of electricity have to compete in bidding processes introduced by the PT. Distributors and large-scale consumers are obliged to enter into long-term purchase agreements, which are made through auctions organised by the Electricity Regulatory Agency (Agência Nacional de Energia Elétrica [ANEEL]). At these auctions, private companies submit bids below a maximum price set by ANEEL. The most competitive ones are then selected. The bidding system is marked by certain shortcomings. Most importantly, sources of energy that have few negative impacts on the environment – biomass, small hydro, solar and wind – compete with each other. Moreover, wind energy in particular requires costly interconnections – which has been a considerable disadvantage for it in bidding processes because it competes therein with centralised systems.

These shortcomings are somewhat compensated for by policies that support renewables. The Support Programme for Alternative Electricity Sources (Programa de Incentivo às Fontes Alternativas de Energia Elétrica, PROINFA) was launched in 2002. It introduced a guaranteed feed-in price for electricity from biomass, small hydro and wind. This price was technology-specific, and reached a maximum 80 per cent of the average final consumer price. It was fixed for 15 years, starting from the commencement of operations. PROINFA set a capacity goal of 3.3 GW, to be generated by the just-mentioned sources. They should account for 10 per cent of total electricity supply by 2022. When the programme was eventually terminated in 2011, it had until then resulted in the deployment of a total capacity of 2.6 GW. Its costs were shared by all electricity users, that through increased prices (Luomo 2014).

Presently, renewable energies are supported by tax exemptions. Several federal states have waivered sales duties for the solar industry. Import taxes on the capital goods used to manufacture photovoltaic equipment and wind turbines have been reduced. In 2015 the Ministry of Mines and Energy launched a programme that aims at turning 2.7 million Brazilian consumers into producers of solar energy by 2030, installing photovoltaic panels on commercial or residential properties. The National Development Bank has become the major funding source for the low-cost financing of renewable energy projects, sometimes partnering with the Inter-American Development Bank.

POLITICO-ECONOMIC CONSTELLATIONS

Brazil's energy policy is driven by a quest for autarky. The motivation for this paradigm, which implies foregoing considerable prospects for regional cooperation, is twofold. First, regional cooperation has proven to be risky, as we explain below. Second, developing Brazil's tremendous domestic potential not only serves the purpose of generating energy. It also means facilitating national economic development and, in the case of hydro, oil and gas resources in the Amazon area, effectively integrating peripheral regions into the national economy.

Non-hydro renewable energies appear to be neglected. Besides the aforementioned measures, there are few policies in place to support solar and wind energy – whose contribution to the national energy matrix remains marginal in spite of their considerable potential. Electrification programmes in the north and north-east of the country include biomass, solar and wind power. Solar-based water-heating systems are subsidised in urban areas, especially through energy-saving programmes. Based on interviews conducted with solar and wind energy companies, Ramos Martins and Bueno Pereira (2011) argue, however, that insufficient government support constitutes the main obstacle to advancing these types of energy. Developing solar and wind energy would require large initial investments to build infrastructure, to develop essential knowledge – for example on microclimates and surface topography – and to train labour. Programmes that serve these purposes do not exist on a sufficient scale. Brazil's economy has just turned from a significant recession to stagnation, while the Temer government – as well as President Bolsonaro now too – aim at austerity. It is, hence, unlikely that additional funds for corresponding programmes will be available any time soon.

As noted, regional cooperation on energy is of little relevance. Events in the first decade of this century demonstrated that relying on neighbouring countries can be highly risky. The price Brazil had to pay for Bolivian natural gas went up to USD 98 per 1,000 cbm in 2003 due to market dynamics. Its average price had been USD 80 when Gasbol was first inaugurated. The price further increased afterwards, too: to USD 116 in 2005 and to USD 159 in 2006 (Sennes and Narciso 2008). In that latter year, the Bolivian government decided to renationalise the country's resources. For Petrobras, this meant renegotiating its contract – which ended up with the company paying much higher royalties. Brazil thus now seeks to reduce its dependence on Bolivian natural gas. The import volume from this neighbouring country is to be kept stable at 30 million cbm a day, whereas domestic production is to expand from 49 million cbm a day in 2010 to 116 million in 2019 according to pre-crisis plans (Ministério de Minas e Energia 2010). A project similar to Gasbol that Brazil sought to carry out with Argentina failed already in its early stages, because of irreconcilable differences over prices (Mares 2006).

On the national scale, politics has also had a strong impact on the oil and gas sector. Petrobras is the dominant player, holding key positions in down-, mid- and upstream activities. The company, founded in 1953 to ensure Brazilian ownership of domestic resources, had a monopoly on activities related to oil and gas up until 1997 – when the government eventually opened the sector up for competition. Since then, more than 50 companies – mostly major international players such as Anadarko, Chevron, Shell, Statoil and Sinopec – have become active in the country, none coming close to Petrobras's relevance though (EIA 2015f). Petrobras still controls 95 per cent of Brazil's crude oil production (upstream dominance). It operates 13 of the 17 refineries that exist in the country (downstream dominance), and is charged with the maintenance of pipelines (midstream dominance). The company is also responsible for the vast majority of domestic natural gas production, and for imports from Bolivia. Petrobras controls the national pipeline network for natural gas, and holds a stake in 21 out of Brazil's 27 state-owned distribution companies (EIA 2015f).

From 2003 onwards, the PT pursued policies on the oil and gas sector that largely reversed the liberal agenda of the preceding government. In 2009, the state-owned company Pre-Salt Petroleum (Pré-Sal Petróleo SA, PPSA) was created. It is responsible for administrating Brazil's pre-salt resources, including the granting of licences. Along with the PPSA came a system of shared contracts, according to which the government not only gains royalties and taxes from the oil and gas sector. The oil produced from pre-salt findings is divided among the contractors and the government. If exploration fails, the contractors will carry the entire financial losses. Legislation passed in 2010 specified that Petrobras had to be the sole operator for Brazil's offshore pre-salt reserves. It had to hold a minimum stake of 30 per cent in all new pre-salt projects. Petrobras could also assume 100 per cent ownership without going through a formal bidding process.

The Temer government has since modified this legislation, allowing Petrobras to reject any new pre-salt project and thus enabling foreign companies to enter into any bidding process. Brazil also had strict local content requirements, which committed companies to source specific percentages of their equipment, services and workforce from within the country. Companies that offered higher percentages of local content had better chances of acquiring the rights to explore an oil block. These ambitious local content requirements have now been eased, so as to ensure foreign investment at a time of a low global oil price (Scholvin et al. 2017b).

Although the government strongly influences the activities of Petrobras, it would be wrong to say that the latter is merely an arm of the government. The company's lines of action, decision-making processes and strategies have reasonable autonomy, as Sennes and Narciso (2008) point out. The government holds 33 per cent of Petrobras's capital and 56 per cent of the firm's preferential shares. Such shares consist of a company stock with dividends that are first paid to shareholders before common stock dividends are paid out. In the event of company bankruptcy, preferred-stock shareholders have the right to be paid out of company assets first. What matters more regarding the relationship of the Brazilian government and Petrobras is that preferential shares do not give the respective shareholders voting rights, however.

Petrobras's central role in the Brazilian oil and gas sector is problematic insofar as the present crisis of the company – which has effectively meant a standstill of its investments – paralyses the whole sector, because of the aforementioned strict legislation on local content. Petrobras is under investigation both in Brazil and in the United States for bribery and money laundering. This multi-billion-dollar scandal started with the arrest of Paulo Roberto Costa, head of refining operations at Petrobras, in March 2014. It then escalated with allegations of government corruption and a kickback scheme being in place, resulting in multiple arrests and the resignation of Petrobras's chief executive officer, Maria das Graças Foster, in February 2015. Leading politicians, including former president da Silva, have been accused of using their influence (and receiving bribes) so that Petrobras signs heavily inflated contracts with private companies. At the apex of the corruption scandal, Petrobras made losses of USD 10 billion in the last quarter of 2016. The company had also made losses in the previous three quarters, albeit smaller ones (Cunningham 2016).

While the investigation is still ongoing, the company's auditor refuses to certify its financial statements – which has kept Petrobras from accessing international capital markets. The coming to light of the corruption scandal and the collapse of the global oil price, events occurring in tandem, have forced the company to undertake a sizeable disinvestment plan. Petrobras (2016) now estimates that its production of oil and gas will reach 3.41 million barrels of oil equivalent a day by 2021, which means a decrease by 30 per cent as compared to previous planning.

ENERGY-RELATED KNOW-HOW

The aforementioned flex-fuel vehicles are a first example of the know-how that shapes Brazil's energy sector. Such vehicles were initially produced in the United States in the 1980s. The technology developed there is hardly applicable in Brazil, because corresponding vehicles in the United States use a special sensor to identify the fuel and adjust the engine. This technology is expensive and impractical for Brazil, where the automobile fleet is dominated by small, low-cost vehicles. Brazilian engineers, hence, figured out how to use sensors already existing in cars, and match their readings against information stored in an on-board computer so as to adjust the engine.

In order to institutionalise research on ethanol production, the National Bioethanol Science and Technology Laboratory (Laboratório Nacional de Ciência e Tecnologia do Bioetanol [CTBE]) was established in 2009. The CTBE aims at enabling economically feasible sugar cane production and other forms of biomass used to generate energy, for example through mechanical processing in agriculture and innovations in plant biology and green chemistry. Brazil has also developed technologies that allow sugar-based alcohol factories to be more productive, by using waste from their production to generate thermoelectric energy – usually at smaller factories connected up to the traditional ones. The sale of electricity corresponds to around 35 per cent of the projected profits in new sugar-alcohol factories.

The most important organisation that deals with energy and innovation is the Research Centre for the Power Sector (Centro de Pesquisas de Energia Elétrica [CEPEL]), managed within the framework of Eletrobrás – the state-owned electricity provider. CEPEL was founded in 1974. Most of its facilities are located in a science and technology park situated on the Ilha do Fundão in Rio de Janeiro. CEPEL's 30 laboratories carry out research on the generation, transmission and distribution of electricity, so as to develop and deploy sustainable technological solutions. The organisation has, furthermore, supported Eletrobrás and the Ministry of Mines and Energy in international initiatives that deal with hydropower and low-carbon technologies.

The development of energy-related technologies has been financed since the 1980s, largely through funds generated by increased electricity prices. In the course of a privatisation wave in the 1990s, concession contracts were introduced containing clauses obliging private companies to invest in energy efficiency and research and development. At the end of that decade, the minimum for corresponding investments by privatised utilities was set at 1 per cent of the net income of each utility. Part of this money is used for projects carried out by the utilities themselves; part of it goes to the CT-Energ, a public fund that finances energy projects, in particular with regard to energy efficiency.

The cases described so far are about adapting existing technologies to the specificities of Brazil, and developing innovative solutions on only a rather small scale. Yet, Brazil plays a leading role globally in technology development for hydropower – and even more so for offshore oil exploration. In 2014 the French company Alstom (2011, 2014) inaugurated a technology centre for hydropower in Taubaté, in the federal state of São Paulo. The investment was driven by market opportunities for hydropower plants capable of adapting to river-flow variation. These plants use so-called Kaplan Turbines, which are to be developed in Taubaté. Brazil today accounts for 45 per cent of the worldwide market for Kaplan Turbines. In other words, the considerable opportunities presented by the Brazilian energy market trigger investments by major transnational companies seeking to develop new technologies in the country.

Hydropower technologies may also be exported. The newspaper *China Daily* reported on 6 May 2017 that there was considerable potential for Brazil to supply technology for large-scale projects – including energy transmission – to China. The two emerging economies have set up the China–Brazil Centre for Climate Change and Innovative Technology for Energy, which brings together experts from their respective leading engineering and research institutions. The centre concentrates on four specific areas: biofuels, CCS, deepwater oil and wind energy. In spite of such endeavours and potential for inter-company technology transfers, China's involvement in the Brazilian energy sector – especially in oil and gas, transmission grids and wind power – is driven ultimately by market opportunities (Husar and Best 2013).

Whereas innovation in hydropower is thus dominated by foreign companies operating in Brazil, Petrobras – as said, the technology leader for pre-salt exploration and extraction – received the OTC Distinguished Achievement Award for Companies, Organisations and Institutions for the third time in 2015. The award is granted every year by the Offshore Technology

Conference, which brings together major business associations and firms involved in the offshore oil and gas sector. The fact that Petrobras has already thrice received the prestigious award confirms the company's outstanding performance. Further to this, Brazil's oil sector has triggered foreign investment aimed at knowledge generation. Several major oil companies, including outstanding service providers such as Baker Hughes and Schlumberger, have invested in the aforementioned science and technology park in Rio de Janeiro, where they cooperate with Petrobras, make their own technologies applicable to operations off the coast of Brazil, and obtain knowledge from Brazilian partners that they may use elsewhere in similar projects (Scholvin et al. 2017a).

CLIMATE POLICY

As Viola and Franchini (2017) explain, Brazil used to be some kind of "climate villain" throughout the 1990s and up until the beginning of this century because of extreme deforestation in the Amazon Basin – which accounted for almost three-quarters of the country's emissions during those years, and made Brazil one of the major global emitters. The government failed to develop a meaningful agenda on climate change mitigation. This changed in the early years of the new millennium, when da Silva became president. Between 2005 and 2010, Brazil made a fundamental shift in terms of climate change mitigation. It was able to bring deforestation under control. The key anti-climate mitigation sector during this period was the oil industry, due to the aforementioned vast pre-salt findings – demonstrating that the PT was more committed to economic development than to environmental protectionism. Viola and Franchini (2017) argue that reformists like the Ministry of Environment already became less influential during Dilma Rousseff's presidency, which displayed a neglect for environmental issues. This stance was aggravated by the subsequent economic crisis.

Especially during the da Silva presidency, Brazil made a commitment to relatively reduce its greenhouse gas emissions. In the context of the UN Climate Change Conference in Copenhagen, it pledged to limit deforestation and promote reforestation so as to cut emissions by between 751 and 772 million tonnes of carbon dioxide equivalent by 2020. Adaptations in the agricultural sector are to reduce emissions by another 50 to 62 million tonnes. Energy efficiency and changes in the national energy matrix are to account for a reduction by 161 to 202 million tonnes, all calculated against a business-as-usual scenario. According to the country's INDC, emissions are to reach 1.2 billion tonnes of carbon dioxide equivalent by 2030. Following a business-as-usual scenario based on emissions in 2005, Brazil would otherwise emit 2.1 billion tonnes by that year. In other words, the emerging economy aims at a reduction therein by 43 per cent. Per capita emissions will decline from 6.5 tonnes of carbon dioxide equivalent in 2012 to 5.4 tonnes thereof 18 years later.

These international commitments are accompanied by various policies on the national scale. Regardless of questions about their sufficiency, energy conservation and efficiency – especially in the industry and transport sectors – are official long-term policy goals. Brazil had introduced corresponding programmes already in 1985 (for electricity) and 1991 (for fossil fuels), respectively. The National Energy Conservation Programme focuses on education and labelling, so as to promote the sensible use of electricity. Still, end-user prices for electricity do not reflect the costs of electricity generation. In some cases, energy conservation and efficiency have been used as a means of industrial policy. For instance in 2012 and 2013, legislation was modified to grant tax incentives to car manufacturers producing domestically and according to high emission standards. These measures – also known by the programme's Portuguese acronym, Inovar – are not motivated by climate change mitigation but rather by industrial development.

The National Energy Efficiency Plan from 2011 proposes incentive mechanisms for a sensible use of energy, better policy coordination and integration, new regulations and further studies. The only quantified objective in the plan is to reduce electricity consumption by 10 per cent by 2030 as compared to a business-as-usual scenario (Ministério de Minas e Energia 2011). The National Plan on Climate Change, published in 2008, specifies that Brazil seek to reduce deforestation by 70 per cent by 2017. Using solar heating for water will be encouraged. A national photovoltaic industry is to be supported. Electricity losses during transmission and distribution are to be reduced, and there is to be an increase of the share of biofuels in the national energy mix. How these objectives will be reached is not explained in depth (Governo Federal 2007).

In 2009, the National Policy on Climate Change (Política Nacional sobre Mudança do Clima [PNMC]) turned into legislation. The PNMC – and many subnational initiatives alike – aims to align socio-economic development with the protection of the climate system. Essentially, this means that the objective of economic growth – which is seen as the path towards socio-economic development – is maintained, but it is to be less energy intensive and based on less hazardous sources thereof. The PNMC states that greenhouse gas emissions are to be reduced by between 36.1 and 38.9 per cent by 2020, as compared to a business-as-usual scenario (Ministério do Meio Ambiente 2010). These figures may sound ambitious, at first. However, they in fact mean that Brazil's energy-related emissions will increase by 56 per cent as compared to 2012 (Luomi 2014).

Sectoral plans have been elaborated as well in the context of the PNMC. One for the manufacturing sector calls for a monitoring, reporting and verification system for industrial emissions. The mining industry is to develop three mitigation programmes: a switch to lower-carbon energy sources, the installation of energy-efficient equipment and the use of new mining technologies. A plan for transport and urban mobility proposes to introduce energy-efficient modes of transport and to upgrade related infrastructure. The emissions generated by freight are to be reduced by switching from road transport to railway lines and waterways.[3] Reducing carbon dioxide emissions from energy generation and transport is not, however, the most important issue for climate change mitigation in Brazil. Eighty-three per cent of Brazil's carbon dioxide emissions derive from agriculture, especially cattle ranching, deforestation and other forms of land-use change. Given the intense deforestation, the World Resources Institute (2014) ranked Brazil as the seventh-biggest emitter of greenhouse gases in the world in 2010. In terms of cumulative emissions between 1990 and 2011, Brazil ranks fifth in a list of the countries most responsible for global warming.[4]

Deforestation is closely linked to Brazil's supposedly environmentally friendly biofuels. Those who support biofuels argue that the plants convert carbon dioxide into oxygen during growth. Hence, they have a neutral climate impact. Because the Amazon Basin is not suitable for growing sugar cane, biofuel production – as long as it is not based on soybeans – does not cause deforestation there. Indeed, sugar cane production concentrates in south and south-east Brazil (Pinguelli, Pires and Muylaert 2009; Rodrigues 2015). However, the cultivation of sugar cane may push livestock farming further into the Amazon Basin – thereby contributing to deforestation in an indirect way, one may suppose.

What is more, the production of ethanol causes emissions in numerous ways: fossil fuels are used to produce fertilisers and other inputs; lorries that transport sugar cane and tractors both run on fossil fuels; the distillation process also requires fossil fuels; and, the same applies to the generation of electricity used for ethanol production too, especially in rural areas where diesel generators are a common source of electricity. Some researchers conclude, therefore, that this seemingly environmentally friendly source of energy in fact generates more greenhouse gas emissions than conventional diesel and petrol do – depending on the crops converted into biofuels, and the native habitats transformed into agricultural land (Fargione et al. 2008; Macedo, Seabra and Silva 2008; Muylaert, Silva and Campos 2009).

Climate Action Tracker, which previously concluded that Brazil's targets are at the least ambitious end of a fair contribution to global mitigation, now argues that these are "insufficient" – meaning not consistent with meeting the goals set by the Paris Agreement. Instead, if all countries contribute to climate change mitigation to the paltry extent that Brazil currently does, global temperatures will increase by 2 to 3°C. Recent developments in energy policy – namely, the Temer government strongly favouring fossil fuels over renewables – as well as the reversal of policies directed against deforestation are the most significant problems faced. Climate Action Tracker criticises the fact that progress in forestry emissions mitigation observed since 2005 has now come to a halt, with deforestation and resulting emissions increases picking up speed once again. Total deforestation increased almost 30 per cent in 2016, which is incompatible with the promises made in Paris. Budget cuts of 50 per cent to the Environment Ministry and 70 per cent to deforestation monitoring authorities – being part of the austerity measures advanced under President Temer – raise concern around the ability to even monitor deforestation. Only the country's current economic crisis works against greater carbon dioxide emissions.[5]

A few weeks before we finalised the manuscript for this book, a far-right candidate won the presidential election based on a loose coalition of agribusiness, evangelicals and the gun lobby. Like his role model Donald Trump, President Bolsonaro is convinced that climate change is not man-made. He has called Brazil's adherence to the Paris Agreement into question, also arguing that the agreement interferes with the state's sovereignty over the Amazon Basin. Bolsonaro is in favour of opening up indigenous areas – currently protected by federal law – to agricultural and commercial use. He railed against "excessive" policing of forests and rural areas during his election campaign. Such policing aims to prevent illegal logging. He also threatened to strip public environmental agencies of their power to impose "fines all over the place", and floated the idea of merging the agriculture and environment ministries – which critics worry would enfeeble environmental protections. After his election victory, Bolsonaro declared that Brazil would back out of hosting the UN Climate Summit in 2019. In one sentence, Brazil's achievements in climate change mitigation and even its modest commitment to such efforts in the coming years are likely to be reversed.

NOTES

1. This information is available online at: www.tps-data-portal.org/all-datasets.
2. Research on the impact of biofuels on food prices has become very broad in recent years, as numerous articles published in the journal *Energy Policy* exemplify. It has not, however, come to a definitive conclusion hereon.
3. These plans are all available online at: www.mma.gov.br/clima/politica-nacional-sobre-mudanca-do-clima/planos-setoriais-de-mitigacao-e-adaptacao.

4. Both rankings count the member states of the EU as one entity. Brazil is, of course, much less relevant if cumulative emissions since the Industrial Revolution are taken into account.

5. This information is available online at: http://climateactiontracker.org/countries/brazil.html.

Chapter 6

Mexico

Mexico's energy sector is based on oil and gas, with increasing natural gas imports and declining domestic oil production. There is considerable potential for solar and wind power, which remains largely untapped. This chapter concentrates on the liberalisation of Mexico's energy sector. We show that it aims at facilitating investment in electricity grids, power stations as well as the down- and upstream oil and gas sectors.

Before coming to this liberalisation, we summarise how Mexico's energy demand is likely to increase considerably, and elaborate on the government's expansion plans for power generation. We argue that Mexico recently passed a critical juncture, and has turned from coal to natural gas and not to renewable energies. Further to this, the chapter shows that this emerging economy is weak in terms of energy-related know-how. It used to be proactive on climate change mitigation and has some sound policies on energy efficiency in place, but the last government shifted back to a focus on economic development – which means the exploitation of both conventional and unconventional oil and gas deposits. Against this backdrop, it is not surprising that Mexico's current policies contradict its own INDC.

MATERIAL CONDITIONS

Mexico's energy consumption is dominated by oil (46 per cent) and natural gas (40 per cent), followed by coal (7 per cent) and hydropower (4 per cent). Non-hydro renewables and nuclear power play marginal roles (EIA 2016b). Natural gas is even more important for electricity generation, reaching a share of 51 per cent therein. Oil and coal are the second- and third-most important sources of electricity, with shares of 19 and 12 per cent respectively. Hydropower (11 per cent) plays a considerable role, followed by biomass, geothermal and nuclear energy (IRENA 2015b). International trade in electricity remains marginal: cross-border grids enable interchanges of 50 MW with Belize, 120 MW with Guatemala and 1,913 MW with the United States (SENER 2016).

Mexico is still one of the largest oil producers in the world, with an output of 2.8 million barrels a day in 2014, but its own domestic resources are depleting. Peak production occurred already in 2004. The output has dropped since then by almost one million barrels a day. Almost 80 per cent of domestic production is consumed in Mexico by now. In 2005, half of it was exported. Also because of the collapse of the global oil price, earnings from the oil sector declined from 30 per cent of the country's export earnings in 2009 to 6 per cent thereof in 2015. What is more, Mexico is a net importer of petroleum products. The emerging economy is not in a position to achieve competitiveness in the downstream sector any time soon because of trade with the United States, where refineries benefit from significant economies of scale and are technically more sophisticated so that they can process the heavy crude oil extracted in Mexico itself (EIA 2016b).

Mexico's own oil production is concentrated in the Bay of Campeche, situated in the Gulf of Mexico. The output of the once-largest oil field, the Cantarell Field, has dropped to around 15 per cent of its former volume. Proven resources amount to a total of 9.7 billion barrels, which puts Mexico at the lower range of the global top 20 – roughly on a level with Angola and Norway. Mexico is believed to possess considerable deepwater reserves. Due to the high costs of development, which result from geological uncertainties

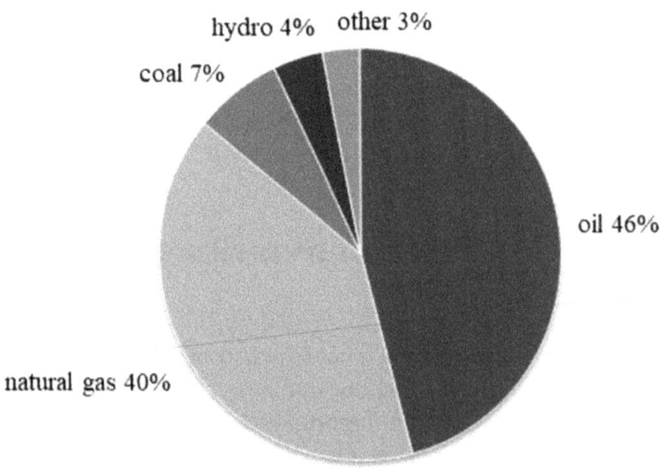

Figure 6.1 Mexico's Primary Energy Consumption by Source. *Source*: EIA (2016b).

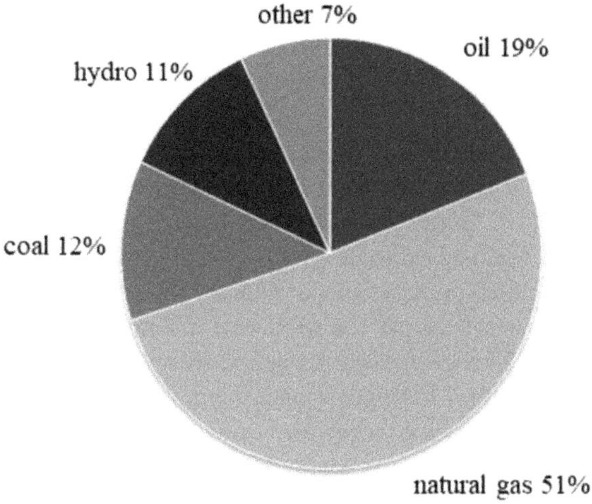

Figure 6.2 Mexico's Electricity Generation by Source. *Source*: IRENA (2015b).

and insufficient infrastructure, these reserves – and also tight oil – have not been exploited yet (EIA 2016b).

The increasing relevance of natural gas is based on relatively cheap imports from the United States. Domestic production, meanwhile, stagnated at the beginning of this decade. It has declined slightly in recent years. Around three-quarters of Mexico's natural gas production stems from associated gas, meaning the variety that is set free during oil extraction. Proven resources reach 433 billion cbm. Even more proficient are the country's shale gas resources, estimated to amount to 15.3 trillion cbm – the sixth-largest in the world. The exploitation of shale gas is hindered by the discontinuity of reservoirs and cheaper imports from the United States. Eighty per cent of Mexico's natural gas imports come from its northern neighbour, equalling 82 million cbm a day in 2015 – up by some 200 per cent since 2010 (EIA 2016b). Imports are likely to increase, as new pipelines are built: in addition to 17 existing cross-border ones, 20 more are currently at different stages of construction (IEA 2016c).

Unlike many of the other emerging economies, Mexico possesses rather miniscule coal reserves. Production increased up until 2011, reaching 9.4 million tonnes of oil equivalent a year, and then dropped to 7 million, which equals about half of the domestic demand (EIA 2016b). Due to the insufficient quality of domestic reserves, about 80 per cent of the coal demand has to be met by imports. Still, three of the country's largest power stations – Petacalco in the province of Guerrero, and Coal I and Coal II in Coahuila

– are coal-fired. Economically recoverable uranium resources are also rather modest, reaching a total of 60,000 tonnes. There is only one nuclear plant in Mexico: the Laguna Verde Power Station, whose two reactor blocks reach a joint capacity of 1.4 GW.

The potential for renewable energies, meanwhile, is considerable. Mexico's geothermal potential is the fifth-largest in the world, according to optimistic studies (EIA 2016b). Gutiérrez Negrín (2012) suggests that a potential of 8.5 GW is a more realistic estimate. Average daily solar irradiation is 5 kWh per square metre in 75 per cent of the country, reaching 7 kWh per square metre near the Cortés Sea and the Gulf of California. Even in the coastal zone of the Gulf of Mexico, which is marked by more clouds, the daily solar irradiation rate averages 4 kWh per square metre (Villicaña Ortiz et al. 2015). The wind power potential is concentrated on the Isthmus of Tehuantepec, one of the windiest places on Earth, where 1,600 turbines are already in operation (EIA 2016b; IEA 2016c). Some studies conclude that north Mexico also yields suitable conditions for wind energy (Hernández Escobedo et al. 2014). No larger wind parks have been built there, though.

According to IRENA (2015a), Mexico holds the potential to increase the share of renewables in its total energy consumption to 21 per cent. Forty-six per cent of its electricity could be generated from renewable sources by 2030, with solar and wind power reaching a joint share of 26 per cent (equalling 60 GW), followed by hydropower (17 GW), geothermal (4.5 GW) and biomass (4 GW). A third of this potential derives from the traditional use of biomass for cooking or heating being replaced by modern forms of renewable energy. Yet, such opportunities have hardly been tapped into. The fleet of 72 state-operated and 28 private hydropower stations is dominated by medium-sized plants with a three-digit maximum output and small plants that reach a capacity of less than 30 MW.

Only three power stations, each built in the 1960s and 1970s and later expanded, are considerably larger: the Chicoasén Dam (2,430 MW), the Infiernillo Dam (1,120 MW) and the Malpaso Dam (1,080 MW). The present maximum output of wind power is 3 GW, followed by geothermal energy (1 GW), bioenergy (780 MW) and solar power (235 MW) (IRENA 2016). The only larger plant in this category is the Cerro Prieto Geothermal Power Station, which reaches a maximum output of 820 MW. In particular, the expansion of wind energy is hindered by the decentralised structure of this source of energy and Mexico's sparse electricity grid. Related to this problem, the emerging economy suffers from high distribution and transmission losses, equivalent to 14 per cent of total electricity generation. The larger part of these is due to non-payment and theft, though (IEA 2016c).

Beyond electricity generation, biofuels have been supported by the government since 2008 – reaching a production capacity of 33,000 litres a day in

2014. Actual deployment is rendered complicated by market size. Only 0.8 per cent of the energy used in the transport sector is based on biofuels. Diesel and petrol prices are expected to increase by 30 to 40 per cent by 2030 – compared to their respective levels in 2010 – because of subsidies being cut. Even under such conditions, biofuels will remain slightly more expensive than conventional ones. Competition with food crops and the consumption of scarce water resources for biofuel production are further concerns (IRENA 2015a).

ENERGY DEMAND AND EXPANSION PLANS

Mexico's electricity demand increased by an annual average of 2.7 per cent from 2005 to 2015. SENER (2016) expects this figure to rise to 3.4 per cent between 2016 and 2030. Peak demand is expected to increase by 3.7 per cent a year, reinforcing the need to expand generation capacities. The IEA (2016c) predicts that Mexico's electricity demand will increase by 85 per cent until 2040, and that the country will need investments of USD 10 billion a year in order to meet this. A study by the consultancy McKinsey (2014b) suggests that energy demand could be 79 per cent higher in 2024 than it was in 2010. The same study concludes that the lack of a cost-efficient and reliable energy supply constitutes the single major challenge for companies operating in Mexico. SENER (2015), similarly, points out that high energy prices are a medium- and long-term risk to competitiveness.

At the same time, the government has set the target of increasing the share of clean energy to 35 per cent by 2024 and to 50 per cent by 2050 (SENER 2016). It will be challenging to achieve this objective because the transport sector is the largest energy consumer, reaching a share of 40 per cent, followed next by the industrial sector (28 per cent) and by buildings (20 per cent). Transportation also accounts for 35 per cent of Mexico's carbon dioxide emissions; power generation for 32 per cent (IEA 2016c). Energy use for transport increased by 66 per cent between 1990 and 2012, driven by the demand for private vehicles and light lorries (Solís and Sheinbaum 2013). The vehicle fleet has grown from 14 million in the year 2000 to more than 30 million today (IEA 2016c). The application of clean energy for transport is, of course, more difficult than it is for electricity generation.

In spite of a rather modest growth in energy consumption in the course of the last two decades, the industrial sector is still energy intensive. The iron and steel industry is first in energy demand among industrial users, accounting for about 14 per cent of the total and a similar share of related carbon dioxide emissions, too. Energy-intensive industries are responsible for 45 per cent of the total industrial energy demand (IEA 2016c). However, steel production in Mexico is more energy efficient than in China and the United States, among other reasons because scrap steel is used, energy-efficient technologies are applied and the share of natural gas as a source of energy for steel plants is relatively high (Rojas Cardenas et al. 2017). Energy intensity in the chemicals industry and in iron production has also declined considerably in recent years (IEA 2016c). In other words, the potential for using energy more efficiently is limited. Mexico will have to change its economy and/or its energy generation fundamentally if it is to mitigate climate change.

In addition to an increasing electricity demand and self-set goals on clean energy, the chances for a fundamental change in Mexico's energy matrix are boosted by the fact that nearly 16 GW of aging generation capacity – out of a total capacity of 65 GW at present – are to be decommissioned by 2030. About 10 GW of the soon-to-be-replaced capacity is conventional thermal (SENER 2014). Yet, it appears that the emerging economy has already passed this critical juncture: most power stations that have been built in recent years are, for purely economic reasons, fired by natural gas. Such power plants are on average 13 years old (IRENA 2015a), suggesting that Mexico will remain on a hydrocarbons path instead of fully switching to renewables. Power generation that relies on natural gas will continue to provide economically competitive, low-cost electricity. This paves the way for politically attractive tariffs, but it may limit investment in renewables – especially because renewable energy projects have usually proven to be uncompetitive (IRENA 2015a).

According to official documents, renewables will, however, complement natural gas. The General Law on Climate Change, adopted in 2012 and last modified in 2016, specifies that 35 per cent of energy generated domestically ought to come from renewable sources by 2024 (Cámara de Diputados 2016). Power stations with a joint maximum output of 57 GW are to be built in the course of the next 15 years, following the Outlook on the Electricity Sector 2016–2030. The yet-to-be-built clean sources are dominated by wind energy (21 per cent), cogeneration (12 per cent) and solar energy (12 per cent).[1] Hydropower (8 per cent) and nuclear energy (7 per cent) account for smaller shares. The origins of the electricity from cogeneration is not specified. Considering that the document refers to combined-cycle coal-fired power stations as a separate category, one can assume that they are not carbon-based, although it may also be that cogeneration refers to all electricity produced in second cycles (including at coal-fired power stations). The remaining 22 GW are to be carbon-based but generated in combined-cycle power plants, which extract energy from waste heat in a second cycle of electricity generation and are thus more efficient than simple-cycle ones (SENER 2016).

Indeed, if Mexico succeeds with its ambitious plans to expand wind energy (and to export the corresponding electricity to the United States), the emerging economy will become one of the fastest-growing wind energy producers in the world (EIA 2016b). Today, only about 7 per cent of Mexico's wind power is generated by the state-owned electricity provider though, indicating a lack of commitment from this side. The remaining wind parks are operated through licensees under self-sufficiency, by small producers and by dependent producer contracts. Self-supply plays a considerable role for renewable energies in general: it accounts for about 11.5 per cent of Mexico's total installed renewables capacity (IRENA 2015).

Most problematically, renewable energies are not encouraged by an established feed-in tariff. To incentivise investment in renewables, the government plans to introduce clean energy certificates, the sale of which will be invested in renewable energy projects. The real market value of these certificates, which came into effect in 2018, is still unclear (IEA 2016c; IRENA 2015). There is by now legislation that establishes the legal aspects and conditions for the use of renewable energies and clean energy technologies. Legislation also seeks to reduce the use of and dependency on fossil fuels. Yet, compared to what some of the other emerging economies have already achieved, Mexico's measures are unimpressive. For instance, the government only in 2017 first allowed the use of petrol with up to 10 per cent of ethanol in the country, except for Guadalajara, Mexico City and Monterrey – where the environmental impact of blended petrol still has to be studied, as the newspaper *El Heraldo* reported on 15 June 2017.

Bringing the aforementioned expansion plans together with the fleet of existing power stations, the capacity of conventional thermal energy is to drop from 12.7 GW today to 2.2 GW by 2030 because of old power stations being switched off. Other forms of carbon-based electricity generation will expand: combined-cycle power generation will increase from 24 GW today to 42.6 GW by 2030. This means that SENER (2016) expects hydrocarbons to dominate electricity generation with a share of 58 per cent, followed by hydropower (11 per cent), wind energy (11 per cent) and nuclear energy (9 per cent). Cogeneration from, as noted, unspecified sources, geothermal and solar energy will play marginal roles.

The implementation of these expansion plans remain somewhat uncertain. A study by Alvarez and Valencia (2016) suggests that the liberalisation of the energy sector, on which we elaborate in the next section, will boost natural gas as a source of electricity and reduce energy costs significantly by replacing more expensive oil fuel – which will also lead to additional growth in the manufacturing sector (and thus to a greater energy demand). Having liberalised its energy sector, the government expects that natural gas production will increase from 161 million cbm at present to 294 million cbm in 2025. Natural gas is meant to meet the bulk of domestic energy demand, whereas oil will be exported because of higher profit margins. According to an interview with SENER's undersecretary of hydrocarbons, oil production

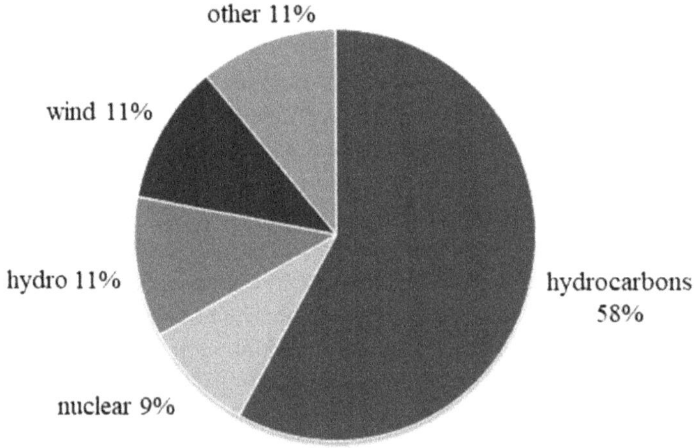

Figure 6.3 Mexico's Future Electricity Generation by Source. *Source*: SENER (2016).

is to rise to 3.5 million barrels a day in 2025 for the same reason (*Mexico Oil & Gas Review* 2014). Following a scenario developed by the IEA (2016c), meanwhile, the share of renewables in electricity generation will increase to 40 per cent by 2040. Almost USD 100 billion will flow into corresponding projects, whereas only USD 40 billion will be invested in fossil fuels.

POLITICO-ECONOMIC CONSTELLATIONS

In 1938, Mexico's oil sector was nationalised and the state-owned company Petróleos Mexicanos, or PEMEX, became a monopolist for down-, mid- and upstream activities. PEMEX was also the sole firm permitted to operate in the mid- and upstream natural gas sectors. Today, it the largest company in Mexico and the seventh-largest oil company in the world. The electricity sector used to be in public hands, too. Despite reforms in the 1990s, private companies could not participate in the transmission and distribution of electricity prior to 2013. Their role in electricity generation was limited to a handful of projects. As a consequence, the state-owned Federal Electricity Commission (Comisión Federal de Electricidad [CFE]) still owns three-quarters of the domestic generation capacity through its subsidiaries. The remaining quarter belongs to about 40 independent suppliers, who operate mostly natural gas-fired power stations. Those that provide electricity to the national grid are small in size. The larger ones power their own production facilities and sell surplus electricity to the CFE, or – in a few exceptional cases – directly to end users.

Against this backdrop, it is not surprising that Mexico did not have an electricity market until recently. Prices were set by the Department of Finance, being mostly subject to political considerations. Electricity used to be heavily subsidised, particularly for agriculture and private households. These subsidies still amounted to 0.6 per cent of government expenditures in 2015 (IEA 2016d); these were sufficient to erode the capital base of the CFE, which had to accept prices below generation costs, but much below their apex in 2011, when subsidies reached almost 2 per cent of the GDP.

In 2013 the Mexican Congress approved a series of constitutional amendments, also known as the Energy Reform. This reform opened up the hydrocarbons and power sectors to private investment. It has not, however, transferred any assets into private hands. The strategy of the Enrique Peña Nieto government (2012–2018) was based on private–public partnerships meant to provide cutting-edge expertise and infrastructure, without ending state ownership. This partial liberalisation is reflected by the priorities stated in the National Energy Strategy 2014–2028: Mexico seeks to maintain public ownership of its own domestic resources. The modernisation of the CFE and PEMEX will not involve any form of privatisation. Yet, at the same time, the Energy Reform is meant to attract investment so as to boost national development (SENER 2015).

As Bonafé, Beltán and Neira (2016) summarise, the Energy Reform serves four objectives: insufficiencies in energy-related infrastructure, especially regarding the electricity grid and natural gas pipelines, are to be overcome; oil and gas production is to increase, in particular through the exploitation of deepwater and unconventional resources; energy supply is to be diversified to avoid over-reliance on natural gas; and, a transition towards a mix of natural gas, nuclear and renewables is to be facilitated. While it appears that the liberalisation will indeed boost investment in oil and gas, a diversification of the energy matrix is rather unlikely. As we explain in this section, the reason for this is that support mechanisms for renewable energies are lacking and this liberalisation will probably trigger, in fact, a hydrocarbon boom.

With regard to hydrocarbons, the Energy Reform consists of five components. First, it calls for the creation of so-called state productive companies. These are public corporations that underwent a process of restructuring – giving them budgetary and operational autonomy – and a series of financial adjustments in order to bring pricing closer to costs. They now participate in wholesale competition. Second, even though the state remains the owner of all hydrocarbon resources it is permitted to pursue their extraction in collaboration with private firms, based on service contracts, production- as well as profit-sharing contracts, and licencing agreements. These arrangements also allow PEMEX to switch to joint ventures in ongoing projects. Third, the

National Hydrocarbon Commission and SENER may conduct bidding rounds and sign contracts with private entities and state productive companies. Fourth, private investors can participate in down- and midstream industries. Corresponding permissions are granted by SENER. Fifth, the Mexican Petroleum Fund has been created to manage and distribute the income that the state derives from the hydrocarbon sector (Alpizar and Rodríguez 2016).

In the electricity sector, transmission and distribution remain integrated in the hands of the CFE – with the subcontracting of private firms and state-owned entities only. Private companies are allowed to negotiate contracts on services and technology provision with the CFE. The resulting partnerships are meant to improve the transmission network. The CFE remains the sole retailer to small customers, providing electricity at rates set by the government. As a wholesale supplier, it has to adapt to competitive markets because other generators and traders are now allowed to do business among themselves. The newly established wholesale electricity market is a spot market, wherein generators, suppliers and qualified customers – that is, mostly large-scale industrial users – gather for buying and selling at real-time marginal costs. The market is based on a merit order of ascending operating costs, under which the power plant running at the lowest expenditure satisfies demand before the next plant is dispatched (Alpizar and Rodríguez 2016).

A centrepiece of the Energy Reform is an auction system for electricity that offers long-term contracts – 15 years for energy and capacity, and 20 years for clean energy certificates. The auctions are technology neutral for clean energy options. The CFE sets out the requirements in terms of electricity, capacity or clean energy certificates. The choice of technology is left to the market (IEA 2016c). As noted, Mexico does not have a feed-in tariff and relies mostly on market-oriented schemes, which are somewhat mediated by different forms of public support for the application of and research on renewable energies. Hence renewable energies compete directly with fossil fuels as far as the latter are considered clean, which is the case for cogeneration. Despite these problems, half of the new generation capacity auctioned for development in 2016 was awarded for renewables (IEA 2016c).

The reform of the electricity sector was driven by high prices for end users due to insufficient investment. Industrial users paid slightly more than their potential competitors from the United States. Private electricity use was on average 25 per cent more expensive than in this neighbouring country. Barriers to private investment also affected transmission and distribution: at the beginning of this decade, 47 per cent of Mexico's transmission lines were more than 20 years old (Alpizar and Rodríguez 2016; SENER 2015). It appears that liberalisation has already paid off, with electricity prices since coming down by 20 to 30 per cent for industry and by 8 to 18 per cent for commerce (Roy and Briones 2016).

The liberalisation of the hydrocarbons sector was equally driven by a lack of investment. The deputy secretary of energy for hydrocarbons under the last government, Aldo Flores Quiroga, pointed out that whereas there are more than 2,300 oil platforms in the territorial waters of the United States, only 46 wells have been drilled off the coast of Mexico (*Oil and Gas Journal*, 8 May 2017). It has to be acknowledged, however, that investment in exploration and extraction increased from USD 4.5 billion in the year 2000 to USD 15 billion in 2012 (Bonafé, Beltán and Neira 2016) – apparently not enough to prevent the dramatic decline in oil production. As noted, Mexico's remaining promising oil and gas resources are either unconventional or deepwater ones. PEMEX does not have the financial and technological capacity to exploit these, considering that the drilling costs in deepwater areas are about 10 times higher than in shallow water ones and 100 times higher than for terrestrial deposits (Alpizar and Rodríguez 2016). The need for foreign investment also affects the downstream sector, as three of PEMEX's refineries – Salamanca, Salina Cruz and Tula – must be reconfigured to eliminate a surplus of high-sulphur oil – which is situated far from the coast, and cannot be economically exported. What is more, whereas natural gas consumption and the related pipeline network are expanding rapidly, Mexico is heading towards a shortage of coking capacity (Lajous 2014).[2]

In order to end PEMEX's monopoly, the Energy Reform reallocated all hydrocarbon resources through a "round zero" process that allowed that company to make initial bids. At the end of this process, PEMEX retained 83 per cent of the proven and probable reserves – the entirety of what the company had requested – and 21 per cent of the prospective resources, which equals two-thirds of its corresponding request (Alpizar and Rodríguez 2016). From now on, PEMEX's privileged access is somewhat conditional. If the state-owned company is interested in a new project, it will have 120 days to prove it possesses the financial, operational and technical capacities to conduct operations. If PEMEX fails to demonstrate sufficient capacities, it will have to relinquish some areas of operation. Liberalising the hydrocarbons sector does not, of course, guarantee foreign investment. The open bidding process subsequent to round zero was not successful because of the low global oil price. It only led to better results when the government offered improved conditions to potential investors (EIA 2016b).

The changes to the hydrocarbons sector go far beyond the reallocation of exploration and extraction rights. In 2016 permits were issued for private companies to import diesel and petrol, effectively ending PEMEX's down- and midstream monopolies. PEMEX is now required to become a profitable corporation. Its revenue is dependent on its own extraction, and the contracts that it enters into. Previous restrictions on private investment in the petrochemicals industry have been lifted. Oil subsidies decreased in 2017 and again in 2018. They will be terminated eventually, so that the price of oil-based energy comes to fully depend on market mechanisms. As of 2018 – after 80 years of government-set prices – natural gas is being sold at market prices, too (IEA 2016c).

However, as a report by the *Oil and Gas Journal* (5 June 2017) stresses, PEMEX continues to be a state agency. There was speculation that the now-deselected Peña Nieto government would, as a next step in its liberalisation agenda, allow the participation of foreign capital in PEMEX. Up to now, the director general of PEMEX has usually been a short-term presidential appointee. The Energy Reform allows the government to auction blocks for oil and gas exploration, but it does not require it to do so. Due to the recent success in the presidential election of Andrés Manuel López Obrador, who has previously argued against liberalisation policies, investment opportunities are likely to shrink going forward. The Energy Reform came about only as a result of a rather unlikely agreement between Mexico's three main political parties during a period of the oil price-induced weakness of PEMEX (Bonafé, Beltán and Neira 2016; IRENA 2015). Violent protests against petrol price hikes by 20 per cent in 2017 (*El Universal*, 1 and 6 January 2017), a follow-up to the Energy Reform, are a reminder of the poor popular acceptance of market liberalisation.

ENERGY-RELATED KNOW-HOW

The Mexican state finances a number of institutes carrying out research on the application of supposedly clean fossil, nuclear and renewable energies. It has also created a fund for energy transition. Most importantly, five Mexican Centres for Energy Innovation have been conceived for bio, geothermal, solar, wave and wind energies, accounting for a total investment of approximately USD 160 million (Bonafé, Beltán and Neira 2016). Mexico is not a technology leader though, which is best demonstrated by the fact that the Energy Reform was largely driven by PEMEX's inability to exploit deepwater and unconventional resources for a lack of both financial and technological capacities. The aforementioned National Energy Strategy 2014–2028 acknowledges that energy-related equipment and technologies are mostly of foreign origin (SENER 2015). Little progress has been made in commercialising clean energy technologies. It remains to be seen whether the government's promise to increase investment in clean energy research to USD 310 million within five years ultimately holds true.

Since the government has set the goal to reduce transmission and distribution losses to 8 per cent by 2024, it will invest in smart grids (IEA 2016c) – this constitutes one potential area for at least applying innovative solutions in Mexico. Smart grids are characterised by the implementation of digital communication, making data flow and information management much more efficient. This allows for demand-side management: the electricity supplier forces all air-conditioning systems to turn off so as to balance short-term demand peaks, for example. Peak curtailment is another option with smart grids. It works through electronic devices that automatically turn off at times of peak demand. Most importantly, smart grids are able to adjust for the short-term fluctuations that mark solar and wind power. They are moreover bi-directional, meaning that they enable distributed generation from photovoltaic panels on building roofs, pumped hydropower and other sources, as well as charging from/to the batteries of electric cars. Traditional grids, meanwhile, are designed for only one-way flows of electricity.

CLIMATE POLICY

For decades, Mexico was an oil-rich country. It still is to some extent. Energy saving and renewable energies received little attention until recently. Emissions increased along with economic growth, slowing down only after 2012. Green growth is now an official objective, one stated in presidential reports. Institutions that coordinate climate change mitigation on the national, state and municipal levels have been established. A climate action plan has been set up by PEMEX (2013, 2016), and the company aims to reduce its carbon dioxide emissions by 25 per cent until 2021 through cogeneration, energy efficiency and less natural gas flaring. The plan entails participation in activities related to adaptation, mitigation and transversal topics.

An increase of natural gas production is to reduce PEMEX's carbon dioxide emissions in relative terms. Related to this, controlling methane leaks in the upstream hydrocarbon sector is critical. Methane emissions are to be reduced by 40 to 45 per cent by 2025, as compared to 2012 levels. Achieving this goal comes at the moderate cost of USD 106 million according to the IEA (2016c), but PEMEX's strained budget makes allocation of capital to such projects challenging – particularly in the short run.

During the presidency of Filipe Calderón (2006–2012), Mexico was highly committed to mitigating climate change and took an active role in bi- and multilateral negotiations. The emerging economy was the second state in the world that translated its ambitions regarding climate change mitigation into law, and one of the first to announce its INDC. Mitigating climate change promised increasing foreign investment, most importantly to change the energy matrix and to boost energy efficiency. Calderón's successor, Peña Nieto, adopted a different approach, declaring in his inauguration speech that he was also seeking to make Mexico benefit from its natural resource endowment. While the furtherance of environmentally friendly energy and corresponding technology transfers was to be facilitated through free trade, the Peña Nieto administration specifically wanted to stimulate development through a revival of the oil sector – that driven by foreign investment (Sosa 2015), as explained above with regard to the Energy Reform.

Some climate-friendly policies were introduced by the Peña Nieto government, including a special action programme with biannual independent monitoring as well as a very low carbon tax on production. A pilot carbon capture and trade programme, announced in 2016, will involve up to 60 companies on a voluntary basis. Fuel-efficiency standards for cars are to become stricter in 2018. No such measures exist for lorries; neither is their introduction planned. This is particularly problematic because heavy goods vehicles are expected to account for more than half of the increase in transport fuel consumption between now and 2040 (IEA 2016c). Energy-efficiency standards for appliances were already introduced in the early 1990s. Energy efficiency in buildings has been encouraged since 2007. A programme launched in 2013 offers loans for ensuring energy efficiency in newly built houses. Starting in 2016, all companies operating in Mexico whose annual emissions equal or exceed 25,000 tonnes of carbon dioxide are obligated to report these emissions to the Secretariat of the Environment and Natural Resources. In addition, the National Emission Registry (Registro Nacional de Emisiones, RENE) has been created to allow companies to register their voluntary mitigation projects. RENE and its database could be used to establish a compensatory carbon tax mechanism. However, these measures lack concrete enforcement, making it difficult to quantify future effects.[3]

In its INDC, Mexico pledges to reduce greenhouse gas emissions by 25 per cent by 2030 (as compared to a business-as-usual scenario) or even 40 per cent if it receives sufficient financial support and technology transfers from the Global North. Climate Action Tracker first rated Mexico's commitment as "medium", meaning that the country's efforts are not fully in line with limiting global warming to less than 2°C. The present assessment thereof, meanwhile, is "insufficient"; the policies in place are only adequate to limit global warming to 2 to 3°C. The predicted emission peak in 2026 is out of sync with the present energy policy. The INDC, furthermore, still remains to be translated into sector-specific action plans, which were in the process of elaboration at the time of writing. The fact that Mexico has included a reduction of black carbon emissions in its climate change mitigation plans – instead of merely specifying the reduction of carbon dioxide and other greenhouse gas emissions – blurs matters further. Reducing black carbon emissions has almost no effect on climate change (although it does generate substantial benefits for human health).[4]

The IEA (2016c) also argues that hitting the aforementioned targets will be difficult with the both implemented and planned energy policies. The power sector alone will absorb the remaining carbon budget. The Energy Reform is likely to lead to a fracking boom,[5] provided that the global oil price increases from its current low. By 2040, Mexico is expected to extract 15 billion cbm of shale gas a year. As explained more thoroughly in chapter 7, the exploitation of unconventional reserves comes along with hazardous emissions and environmental risks unrelated to climate change – especially the extensive use and likely contamination of water. Also due to liberalisation, Mexico's oil output will increase to 3.4 million barrels a day in 2040 – mostly based on the exploitation of deepwater and tight oil reserves. As far as oil is exported, it does not appear among Mexico's own carbon emissions – but such a revival of oil extraction obviously means greater emissions worldwide.

Still, the Energy Reform does somewhat contribute to climate change mitigation because private investment in natural gas will lead to a decline of the share of oil in the country's energy matrix, thus reducing its carbon dioxide intensity. Following this logic, the government considers hydrocarbons that are burnt at power stations equipped with CCS technology and cogeneration to be clean energy. In consequence, the focus of energy policy shifts away from renewables. New hydrocarbon-based power stations – even if equipped for cogeneration – mean more carbon dioxide emissions. One might, of course, expect the liberalised energy sector to attract considerable investment in solar and wind energy. However, as explained above, natural gas has proven more competitive than renewables and Mexico currently lacks sufficient support mechanisms for any fundamental change in its energy matrix.

NOTES

1. Cogeneration refers to the generation of electricity and heat at the same time. Such plants recover otherwise wasted thermal energy.
2. Coking is the conversion under extreme heat of natural gas (or oil) into a combination of products for further processing.
3. The registry's website suggests, however, a serious lack of commitment on the part of private companies; available online at: www.gob.mx/semarnat/acciones-y-programas/registro-nacional-de-emisiones-rene.
4. This information is available online at: http://climateactiontracker.org/countries/mexico.html.
5. Recent statements by President López Obrador suggest that hydraulic fracturing may be banned in Mexico (*Excélsior*, 5 and 19 October 2018).

Chapter 7

South Africa

This chapter stresses the heavy dependence of South Africa on coal. The enormous potential for solar and wind power remains, meanwhile, almost completely untapped. The dominance of coal translates into politico-economic constellations – the interest groups in favour of an energy-intensive, hydrocarbon-based model who shape South Africa's energy policy – and into energy-related know-how, with considerable efforts being made to implement CCS. We analyse the expansion plans for power generation, pointing out that they are outdated because of new resources being available in South Africa and its near abroad and due to a remarkable decline in the costs of renewables.

Nonetheless coal will continue to play a major role in South Africa's energy mix, along with natural gas and, arguably, solar and wind power. Further to that, we explain the political problems that account for an inefficient and unreliable electricity sector, especially the difficult situation of Eskom – the state-owned provider. We also elaborate how corruption has influenced the country's energy policy. With regard to climate change mitigation, the South African government considers economic development to be its priority. Carbon dioxide emissions will thus increase. The expansion of renewable energies has been delayed; so have a carbon tax as well as entity- and sector-specific emission budgets.

MATERIAL CONDITIONS

South Africa's energy sector is based on coal, which accounts for 72 per cent of the energy consumed in the emerging economy and for 94 per cent of the electricity generated by Eskom. The reason for this is that South Africa is exceptionally poor in oil and gas, while it possesses 3.5 per cent of the world's coal reserves. Unlike most Western countries, South Africa did not go through a transition from coal to oil in the 1950s. Today, it accounts for 3.3 per cent of the world's annual coal production and for 2.6 per cent of the world's annual consumption thereof too: 260 million and 205 million tonnes respectively in 2014 (EIA 2015g). There are 12 coal-fired power stations in the country that reach individual maximum outputs of 1.2 to 4 GW. Most of these will be decommissioned in the 2020s and 2030s, meaning that South Africa is fast approaching a critical juncture in its energy policy.

Oil, which is imported, is the second-most important energy source, with a share of 22 per cent. This high share results from the transport sector. Natural gas and nuclear energy play marginal roles, with a share of 3 per cent each. Infrastructure for LNG being practically non-existent, open-cycle gas turbine plants in the Western Cape are run on diesel instead, which is extremely expensive. Renewable energies and hydropower are negligible, accounting for less than 1 per cent of total energy consumption (EIA 2015g).

South Africa's most important coal mines are located in the north-east of the country; so are most coal-fired power stations. The Johannesburg–Pretoria agglomeration is relatively close, but other large cities are distant, complicating energy transmission. Bringing coal to power stations has also become challenging. Coal mines in Mpumalanga are depleting. The resources have to be brought in from Limpopo, largely by lorry because railway infrastructure is insufficient. Transport costs have reached 17 per cent of what Eskom pays for coal (Styan 2015). What is more, the news agency Reuters reported on 11 March 2009 that, according to Eskom's own calculations of that same year, South Africa would need 40 new coal mines until 2020 to meet the projected domestic demand. Opening so many new coal mines is highly unlikely: black South African entrepreneurs lack capital, but have to hold 51 per cent of the shares of the mines with which Eskom signs deals because of existing legislation on economic transition. Further to this, increasing global demand for coal – in particular by China and India – involves the risk that even its low-quality variety (considered unsuitable for export) will become more expensive as soon as Eskom's long-term supply contracts run out.

There are considerable shale gas deposits in the Karoo, a semi-arid region east of Cape Town. They amount to 11 trillion cbm, according to optimist projections by the EIA (2013b). If this were true, the Karoo would host the eighth-largest shale gas deposit worldwide. However, David van der Spuy, an expert from the Petroleum Agency of South Africa, declared to the Energy Portfolio Committee in parliament that there are probably only 1.1 trillion cbm of recoverable resources (*Business Day*, 21 February 2014). International companies such as Shell have shown great interest in exploiting these resources, but the low oil and gas price and unclear legislation – the

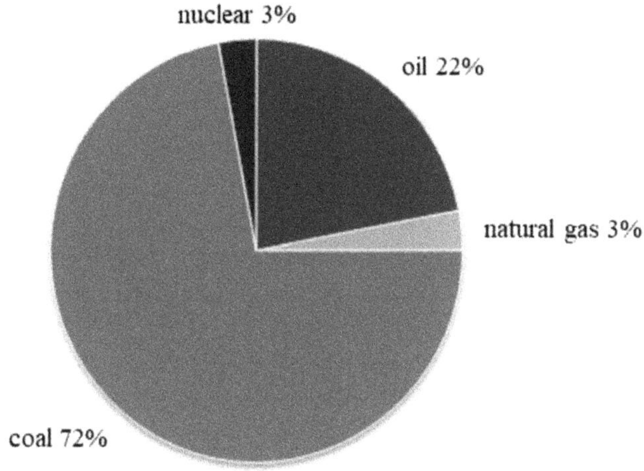

Figure 7.1 South Africa's Primary Energy Consumption by Source. *Source*: EIA (2015g).

government has drafted a law that would allow it to nationalise investments made by foreign companies in the oil and gas sector – have brought exploration to a standstill. Other challenges to exploiting shale gas in the Karoo range from environmental risks, which are reinforced by the insufficient capacities of South Africa to deal with hazardous waste produced in the course of hydraulic fracturing, to the lack of an infrastructure to bring water to the extraction sites – and natural gas from there to consumers (Fig and Scholvin 2015).

What is more, shale gas is a marginal resource. This means that it can be exploited only at times of high oil and gas prices. It depends on an existing infrastructure, especially pipelines and storage facilities. Arguably, tax incentives are necessary for shale gas exploitation to become economically viable. All these factors were present during the North American boom. As said, South Africa lacks an infrastructure for natural gas. Oil and gas prices are relatively low by now, with conventional resources being available in relatively close proximity: namely, in Angola and Mozambique. It is doubtful that the South African government would be willing and able to grant international companies and their local partners tax incentives for exploiting the Karoo's resources.

The Karoo, moreover, hosts vast quantities of uranium. Peninsula Energy, an Australian explorer and producer, speculates that these resources might be 10 times larger than the 26,000 tonnes it has already found there. The company aims to start mining in 2019 or 2020, at a rate of almost 1,400 tonnes a year (*Bloomberg News*, 14 February 2016). South Africa as a whole possesses about 7 per cent of the world's proven uranium reserves, and accounts for 1.1 per cent of global production. At present, uranium is mined in the Witwatersrand Basin as a by-product of copper and gold. South Africa's only nuclear power station – Koeberg, near Cape Town – can generate up to 1.8 GW. At least up until the end of the Jacob Zuma presidency (2009–2018), it appeared that the domestic resources would play a key role in an expansion of the nuclear sector for political reasons (more on this later).

Natural conditions for hydropower are rather limited, given that large parts of the country are semi-arid. Rainfall varies considerably, making it almost impossible for hydropower stations to provide a stable input to the energy grid. Large rivers are few in number. The potential for small-scale hydropower, which could be exploited easily because many dams exist in South Africa, is merely 247 MW (Barta 2011). Presently, Eskom operates two conventional hydropower stations that reach significant output: they are both located on the Orange River, at the Gariep Dam (360 MW) and at the Vanderkloof Dam (240 MW). Two larger pumped storage schemes – Drakensberg (1 GW) and Palmiet (400 MW) – are currently operational. The Ingula Pumped Storage Scheme (1.3 GW) went fully online in 2017. With regard to hydropower, regional cooperation matters more than domestic potential, as we explain below.

Biomass use in South Africa is mostly based on inefficient harvesting from woodlands. More efficient forms would be the use of waste fibre from sugar cane processing and waste wood from the plantation-based pulp industry, both concentrated in KwaZulu-Natal. Agricultural and garden waste could also be used. According to a study by the Council for Scientific and Industrial Research, more than 80 per cent of South Africa's territory has sufficient natural conditions for economically viable wind farms, reaching annual load factors of more than 30 per cent. The study concludes that a widespread wind fleet's fluctuations – within intervals of 15 minutes – are only about 4 per cent of the installed capacity, meaning that this source of energy would be fairly reliable (Knorr et al. 2016). In 2016, wind parks only reached, however, a joint maximum output of slightly less than 1.4 GW.

Natural conditions for solar energy are very good. Seasonality for solar radiation is low. Most parts of South Africa receive more than 2,500 hours of sunshine a year – one of the highest irradiations worldwide. The radiation measures from the best sites in the Northern Cape are more than 30 per cent higher than for the best ones in Spain. Other locations across South Africa are also highly suitable (Fluri and Von Backström 2009). What is more, from a purely economic perspective, it does not make sense to expand the electricity grid to many rural areas. Doing so would be expensive, due to long distances and relatively few users. Rural areas in South Africa are typically poor, and households there cannot pay for electricity in many cases. It would, therefore, be better to equip these households with solar panels and batteries instead.

The aforementioned lack of oil has not only had an impact on the generation of electricity. In the early 1950s, the apartheid regime launched a coal-to-liquids programme in order to guarantee energy security in a more-and-more hostile international environment. It was later joined by a natural gas-to-liquids programme. Hence, coal and natural gas play a major role in fuelling the transport sector today. The company Sasol, which was founded by the government in 1950, has become the largest synthetic fuel producer in the world. Fuel synthesised from coal and natural gas meets 86 per cent of the yearly demand in South Africa. Sasol's petrochemical industries alone are the second-largest coal consumer in the country, just behind electricity generation and ahead of metallurgical industries (EIA 2015g).

Being heavily dependent on coal leads to high carbon dioxide emissions. As shown in the introductory chapter, South Africa is among the top-20 carbon dioxide emitters worldwide – even though the country has a relatively low population size of about 50 million people. Its annual per capita emissions are comparable to industrialised Western countries. Sasol alone emits almost 72 million tonnes of carbon dioxide a year. Its Secunda coal-to-liquids plant is the largest point source of greenhouse gas emissions in the world. A new such plant, which has the government's backing, is to be built in Limpopo. This is highly problematic, because liquid fuels converted from coal are about 60 per cent more greenhouse gas-intensive than conventional diesel and petrol are. Coping with South Africa's tremendous emissions depends on CCS technologies being applied, Eskom, Sasol and the government argue.

In addition to this, the emerging economy suffers from an economic structure that limits its options in shifting towards an energy policy that would reflect the necessities imposed by climate change. The industrial sector contributes 31 per cent to the GDP, and heavy industries account for about 60 per cent of electricity use. Car assembly, chemicals and metalwork – especially aluminium, iron and steel – are key branches. Energy-conversion efficiencies of these industries are all relatively low in South Africa (Nkomo 2005), partly because low-cost electricity has made it almost irrelevant to think about energy efficiency. Nonetheless, overall energy intensity has decreased since the 1990s. A recent assessment concluded that energy intensity plays the main role in change in South Africa's energy-related carbon dioxide emissions, followed by the dominance of fossil fuels (Zhang, Dai and Song 2015).

ENERGY DEMAND AND EXPANSION PLANS

The future of South Africa's energy sector is outlined by the Integrated Resource Plan, published in 2011. This is now outdated, because economic development since has been less impressive than expected. Experts from the Energy Research Centre (2013) at the University of Cape Town argue that peak demand in 2030 will be rather 61 GW and not 89 GW, as originally predicted at the beginning of this decade. Vast natural gas resources have been found off the coast of Mozambique – they could be imported – and the costs of renewable energies have declined. Still, the Integrated Resource Plan remains the basic point of reference. It calls for a diversification and expansion of electricity supply. Forty-two per cent of the additional generation capacity by 2030 – 17.8 GW – shall be solar and wind energy. Nuclear energy is to account for 23 per cent of the new capacities, or 9.6 GW; coal for another 15 per cent, or 6.3 GW. Hydropower and natural gas hardly play minor roles (Department of Energy 2011).

The 2013 update of the Integrated Resource Plan points at lower-than-expected GDP growth and a shift away from energy-intensive industries. According to these projections, the version from 2011 overestimates future electricity demand by at least 6.6 GW. Hence, the decision to invest in nuclear energy should be postponed. Nuclear baseload capacity will not be required until after 2025 or, following lower-demand scenarios, at the earliest in 2035. Alternatives such as imported hydropower, imported natural gas and domestic shale gas should be explored before investing in nuclear energy, the authors of the updated plan suggest (Department of Energy 2013). However, the 2013 Integrated Resource Plan was never ratified by parliament – meaning that it does not serve as a policy framework. Presently, a new version of the plan is being discussed – thus at least recognising that electricity demand and prices have changed. If this plan does replace the 2011 version, new generation capacities built until 2030 will consist of coal (5.3 GW), natural gas (12.7 GW), solar (4.7 GW) and wind energy (6.4 GW) (Department of Energy 2016).[1]

An alternative plan developed by the Energy Research Centre (2013) suggests that South Africa does not need major investments in new capacities before 2025. From that year to 2030, meanwhile, it would be sound to concentrate on natural gas in combination with imported hydropower as well as domestic solar and wind power. No additional nuclear power should come online prior to 2040, because low-emission alternatives can be built at a cheaper price – that with shorter lead times and in smaller increments, thus reducing the risk of an overbuild. Another report by the Energy Research Centre (2015) concludes that a commitment to the nuclear programme as specified in the 2011 Integrated Resource Plan would lead to electricity prices 20 per cent higher than necessary in 2040, as compared to a more flexible energy policy.[2]

Per all plans, coal will remain vital. The Integrated Energy Plan from 2016 even states that the relevance of coal as a baseload source will increase if natural gas does not prove to be cost-competitive (Department of Energy 2016). Kusile and Medupi – two giant coal-fired power stations presently under construction – will reach a maximum output of 4.8 GW each. They will be among the top-10 power stations of their kind worldwide, jointly consuming around 34 million tonnes of coal and emitting 60 million tonnes of carbon dioxide between them each year (Blignaut 2012). Two other giant coal-fired power stations have been discussed under the project names "Coal 3" and "Coal 4". The most ambitious, but by now outdated, announcements suggest that each plant will reach a maximum output of 5.4 GW (Eskom 2008).

Critics argue that Coal 3 and Coal 4 are not about generating electricity at competitive costs. There will not even be a need for the electricity that these power stations produce, should they go ahead (*Business Report*, 14 March 2016). If built, they would rather ensure that coal mining remains viable in the Waterberg area – which contains 40 per cent of South Africa's remaining coal reserves, and is of key interest to politically connected businesspeople (*Times Live*, 19 October 2017). Because of the limited financial capacities of Eskom and indeed the government, it is unlikely that Coal 3 and Coal 4 will become reality. As an alternative, independent power producers could invest in coal-fired power stations, providing new capacities of 2.5 GW and selling the electricity to Eskom. Such plans remain at the stage of only speculation (Styan 2015). South Africa does not have any concrete planning for coal-fired power stations after Kusile.

The nuclear build-up is not moving ahead, either. Regardless of the unclear timing, the key argument for nuclear energy is that it is very reliable and thus compensates for supposedly barely reliable renewable ones. At least according to calculations by the Department of Energy (2011), nuclear energy is also relatively cheap. Yet, South Africa does not have a plan in place for dealing with high-level nuclear waste. Until today, this sort of waste is stored in the cooling ponds of the Koeberg Power Station. Building nuclear power stations is, moreover, a time-consuming endeavour. If the bidding for contracts had started in 2012, the first new nuclear power stations would have become operational in 2024 or 2025, as Dipuo Peters, the minister of energy at that time, expected (*Engineering News*, 15 September 2011). So far, not a single contract to build a nuclear power station has been signed.

What is more, even if one takes into consideration that energy demand has increased (and will continue to do so) more slowly than expected, which means that 4.8 GW of nuclear power instead of 9.6 GW might be sufficient, it is difficult to believe that Eskom will be able to finance this programme; in particular since Standard & Poor downgraded Eskom's credit rating to BB in 2015, meaning "non-investment grade speculative". The government does not have sufficient financial capacities either, and it appears that this is a major reason why the commissioning of the first new nuclear power station may be postponed to 2037 (according to the presently discussed Integrated Resource Plan [Department of Energy 2016]). Such a postponement would be tantamount to ending the programme. At the time of this writing, there was speculation that South Africa's new president, Cyril Ramaphosa, had sent a special envoy to Russia to cancel the nuclear deal supposedly signed by his predecessor, Zuma, and Vladimir Putin (*Mail and Guardian*, 18 May 2018).

The expansion of renewable energy is carried out by independent power producers. They participate in bidding processes in the context of the Renewable Energy Independent Power Producer Procurement Programme (REIPPPP). The government decides how many MW will be put on auction. It also specifies a maximum price per kWh for each source of electricity and takes socio-economic criteria – for instance the contribution of the proposed projects to local economic development – into consideration. Eskom then signs power-purchase agreements of 20 years duration with the successful bidders. A total of 92 projects were approved between 2011 and 2016. They will soon generate slightly more than 6.3 GW of almost exclusively solar and wind power (Eberhard and Naude 2016).

The REIPPPP has been remarkably successful as costs have declined: whereas they were much higher than those of electricity generated by the state-owned electricity provider in the first three bidding rounds, they fell below Eskom's in 2016 (*Business Day*, 21 July 2016; for a more thorough assessment of the REIPPPP, see Eberhard, Kolker and Leigland 2014). Since 2010, meaning when the Integrated Resource Plan that has officially guided South Africa's energy policy until today was first drafted, prices for photovoltaic panels have gone down from ZAR 3.65 per kWh to ZAR 0.60. The costs of wind energy have dropped from ZAR 1.52 per kWh to ZAR 0.62 over the same period. A report by the Council for Scientific and Industrial Research points out that South Africa could run 70 per cent of its energy grid by a mix of solar, wind and peaking power, the latter being based on natural gas and pumped storage schemes. This would be cheaper than a mix of coal and nuclear energy (Wright et al. 2016).

Apart from that, metropolitan areas such as Cape Town and Nelson Mandela Bay have rolled out ambitious plans to equip city-owned buildings and private households with solar water-heaters. This way, the demand for electricity from the grid – and also the need for its expansion and maintenance – is reduced. When the 2011 Integrated Resource Plan was drafted, Eskom calculated that the theoretical potential of solar-based water heating equals 18 per cent of South Africa's total coal-generated electricity. Given that not all buildings can be equipped with solar water-heaters, the technical and economic potential of the latter is expected to remain at a respective 52 and 18 per cent of the theoretical potential by 2030 (Edkins, Marquard and Winkler 2010).

Declining prices for renewable energies and the aforementioned enormous natural potential for solar and wind energy have led some scholars to very optimistic predictions on a coming – or, indeed, already ongoing – energy transition. Bischof-Niemz and Creamer (2018), for instance, develop a road map towards a renewables-based energy matrix that seemingly meets all of South Africa's objectives, not only providing electricity at the lowest cost but also creating more jobs than any of the hydrocarbon- or nuclear-based alternatives under consideration. According to these two authors, a decarbonised system will power a competitive industrial economy, drive an electric-mobility revolution and in the long run create new export opportunities. We acknowledge that renewable energies offer considerable prospects, and that South Africa must pursue a corresponding path for economic and environmental reasons. However, throughout this chapter we elaborate on major obstacles to such an energy transition – ranging from the aforementioned material reality of South Africa's power-generation capacities to the politico-economic constellations that cement in place the hydrocarbon path. Recognising these obstacles, and the resulting limits to the decarbonisation potential of the South African energy matrix, reflects an inconvenient reality that tends to be forgotten by some.

A first problem for the expansion of renewables is that South Africa is presently marked by an oversupply of electricity, not by the insufficient capacities that explain the expansion plans already summarised. As a consequence, Eskom has not implemented the fifth round of the REIPPPP – practically keeping potential providers of renewable energy out of the grid. This fifth round should have taken place in 2016, but at the time of writing it remains unclear when (or whether) it will go ahead. Eskom has also refused to sign outstanding power purchase agreements from the fourth round (*Business Report*, 18 December 2017; *Daily Maverick*, 7 August 2017).

Further challenges for the expansion of renewable energies could result from local content legislation, however. South Africa demands up to 45 per cent of local content in the wind energy industry so as to boost domestic technology-intensive activities. In practice, local content legislation has led to domestic production of low-tech components such as towers, which are difficult to transport. Researchers fear – on account of negative experiences made in Brazil – that the price of wind energy is likely to increase. Their implementation may slow down. South Africa's relatively small domestic wind energy market and insufficient technological capacities reinforce these concerns (Rennkamp and Fortes Westin 2013). They also apply to biomass and solar energy.

The tremendous expansion of generation capacities according to the 2011 Integrated Resource Plan is meant to end the blackouts that have frequently hampered the South African economy. In January 2008, Eskom had to demand from its largest clients – mostly gold and platinum miners – the shutting down of operations and reduction of the use of electricity to minimum levels. Palladium and platinum hit record prices, because South Africa normally supplies 30 and 85 per cent respectively of these metals to world markets (*Business Day*, 22 February 2008). Load shedding had to be imposed again from time to time in both 2014 and 2015 due to high electricity demand, short-term fuel supply deficits and accidents – most critically, when a storage silo collapsed at the coal-fired Majuba Plant, which provides about 10 per cent of South Africa's electricity (*Mail and Guardian*, 4 November 2014). Such events boost the case made by those in favour of rapidly building more power stations, even though South Africa now appears to suffer from overcapacities, which generate unnecessary costs for Eskom (*Business Day*, 6 January 2017).

Against this backdrop, it is not surprising that electricity prices have increased considerably since the 2008 energy crisis: from an average price of just ZAR 0.18 per kWh in 2006/2007 to ZAR 0.82 in 2016/2017 (Eskom n.d.). Electricity prices are set by the National Energy Regulator of South Africa (NERSA), and at least partly reflect Eskom's calculations. In recent years, NERSA has granted much lower tariff hikes than demanded by Eskom (Fin24, 29 June 2015; *Mail and Guardian*, 28 February 2013). For the financial year 2018/2019, the regulatory has granted a price increase by 5.23 per cent instead of 19.9 per cent, as demanded by the electricity provider (*Times Live*, 15 December 2017).

South Africa might face an even worse situation in the near future. As said, numerous coal-fired power stations have to be decommissioned in the next two decades. A crisis in the short run may ensue from poor maintenance, which has already reduced the availability of power stations to 73 per cent – meaning that 11.3 GW out of the current capacity of about 42 GW is unavailable at any given time (Eskom 2015). Only 49 of Eskom's 121 generating units are in a healthy state; 26 are in a poor and 32 in a critical condition. Poor maintenance is a consequence of power stations not being taken off the grid because doing so would cause load shedding. With availability declining, the opportunities to take power stations off the grid for maintenance seemingly decline as well (Styan 2015).

Policy debates on South Africa's energy sector, especially electricity, are much more influenced by the prospects of regional cooperation than in the other emerging economies. In a famous *Foreign Affairs* article published in 1993, Nelson Mandela suggested that there was considerable potential for cooperation on energy in Southern Africa. The White Paper on Energy from 1998 states that South Africa should make its energy policy compatible with its fellow members of the Southern African Development Community (SADC). Three years prior to the formulation of this policy paradigm, the power utilities of the continental member states of SADC established the Southern African Power Pool (SAPP). The SAPP facilitates inter-firm cooperative arrangements. Its members have established a day-ahead market and a short-term energy market. These enable the trading of surplus energy. The SAPP has pushed for the liberalisation of state-controlled energy markets. During the Mbeki presidency (1999–2008), Eskom capitalised on the resulting opportunities and became a major investor in the SAPP countries and beyond. The company lobbied for making energy projects a priority of the New Partnership for Africa's Development (Greenberg 2009).[3]

Before the 2008 energy crisis, regional cooperation served, first of all, political purposes. Rather than advancing economically sound projects, Eskom and its partners talked with each other for the mere sake of having such meetings (Scholvin 2015). The one major successful project of regional cooperation on electricity is the hydropower station at the Cahora Bassa Dam on the Zambezi River, in Mozambique. South Africa's economic hub Gauteng has been connected to this power station since 1979. For a few years, 7 per cent of the electricity used in South Africa came from Cahora Bassa. The power station did not, however, operate regularly from 1981 onwards because of the proxy war raging in Mozambique. Regular generation of electricity at Cahora Bassa resumed in 1998, with South Africa again being the main consumer. At present, regional countries purchase electricity from South Africa instead of supplying it to the emerging economy while, at least since the 2008 energy crisis, the preference of Eskom and the government has been to expand generation capacities at home (Scholvin 2014a, 2014b).

Several opportunities for purchasing electricity from neighbouring countries have been foregone: Eskom and its Namibian counterpart, NamPower, almost signed an agreement on the Kudu Project in 2005. NamPower still seeks to build this gas-fired power station, exporting 100 MW to South Africa; prospects are dim however, as a Namibian energy expert confirmed in a personal interview in August 2016. Eskom has also decided not to commit itself to buying large amounts of electricity from Botswana, where the 3 GW Mamabula Coal Power Station would otherwise have been built for this purpose (*Business Report*, 11 August 2009; *Mmegi Online*, 16 October 2009).

Another project that hardly moved forward is the Grand Inga Dam in Congo-Kinshasa, which may – one day – generate up to 40 GW, including 2.5 GW for use in South Africa (Maupin 2015). Even Ethiopia might, at least according to the visions of its government, provide electricity to other African countries – including South Africa (Cuesta Fernández 2015). A major obstacle to all these projects being realised is the limited financial and technical capabilities of the respective regional countries. Some of them are also politically unstable, especially Congo-Kinshasa and Zimbabwe – the latter, nevertheless, being crucial as a transit country. In addition to that, political negotiations become complicated – with numerous power utilities and governments involved. These problems result not only from scarcely compatible interests; they are also due to language barriers and differences in legal systems (Scholvin 2015).

Beyond electricity, the South African government has considered regional cooperation on oil and gas part of its strategy of economic development. Refining oil for the SADC market and exploiting natural gas fields in Mozambique and Namibia are mentioned as objectives in the White Paper on Energy. At least Mozambique has become a supplier of natural gas to South Africa. Because of imports from that country, the share of natural gas in South Africa's primary energy supply increased from 1.5 to 4 per cent at the beginning of this decade (Norbrook 2012). The Integrated Energy Plan also calls for cooperation with neighbouring countries – Angola, Mozambique and Tanzania, specifically – so as to jointly use natural gas (Department of Energy 2011).

POLITICO-ECONOMIC CONSTELLATIONS

South Africa's energy policy is strongly influenced by deeply held convictions about the relationship between economic growth, socio-economic development and energy consumption. The Growth, Employment and Redistribution (GEAR) strategy, launched in 1996, demonstrates that the African National Congress (ANC) saw (and still sees) economic growth as the path towards socio-economic progress. The Accelerated and Shared Growth Initiative (ASGISA), which replaced the GEAR strategy in 2004, adheres to the same logic. Both documents set annual GDP growth rates of 4.5 to 6 per cent as targets (Department of Finance 1996; Presidency 2005). The New Growth Path, put forward by the Economic Development Department in 2011, aims at five million new jobs by 2020. To meet this objective, annual GDP growth rates of 4 to 7 per cent are considered necessary (Economic Development Department 2010).

The ANC's policies have had positive effects for previously disadvantaged citizens in terms of access to electricity. In the course of the Integrated National Electrification Programme, 2.8 million households were connected to the grid in the second half of the 1990s. Presently, electrification stands at 85 per cent. Eleven to 12 million households will most likely have access to electricity by 2025 (Ziramba 2008). In the context of its pro-poor policy, the ANC has decided to supply every poor household with 50 kWh a month for free. In comparison to the industrial and mining sectors, consumption by poor citizens remains marginal. The only arguably negative impact of the expansion of the grid is that construction work absorbs capacities of Eskom that could, otherwise, be used to increase the overall supply.

Yet, post-apartheid politics have also had negative effects on energy security. The first issue that matters in this regard is load shedding. This phenomenon was to be avoided at any costs during the 2010 Football World Cup. Shortly before elections, especially local ones, Eskom finds itself under pressure from politicians not to cut off voters. The consequence is that Eskom has kept power stations running, even when they should have been taken off the grid for maintenance. Another political issue that affects energy security relates to the respective roles of Eskom and municipalities. Fifty-five per cent of all electricity is sold to end users by municipalities; only 45 per cent directly by Eskom. This has allowed municipalities to finance their budgets by increasing electricity prices. Even worse, many are in debt to Eskom, having received electricity – and usually also the payments from end users – but without then paying the state provider. Related to this, municipalities have to maintain local grids, the ones that connect end users to larger transmission lines. Because of financial shortfalls many do not, however, carry out sufficient maintenance, let alone renew ageing grids.

On top of non-paying municipalities, there are non-paying end users, too. Electricity theft is not a crime in South Africa. People who construct unauthorised connections to the public grid are arrested on charges of trespassing and vandalism, when at all. In addition to poverty, electricity theft results from many ANC supporters being convinced that electricity should be free of charge. For example, in Soweto – an ANC stronghold west of Johannesburg – only 16 per cent of residential users pay their electricity bills (Eskom 2015). Engineering analyst Chris Yelland (2008) estimates that about 10 per cent of the national electricity use was stolen in 2008. This is not only problematic because of lost revenues. If stolen electricity were available to paying customers, load shedding would be much less of a problem – likely, in fact, not even one at all – and the economic losses incurred due to it – namely, factories and mines not operating, and potential investors avoiding South Africa – would not occur.

Eskom, being a state-owned enterprise, has also been affected by politics in a more direct way. Founded in 1923 to generate cheap electricity for an industrialising country, Eskom operated for half a century without much political interference – being run by technocrats, and using its own income to finance the expansion of its capacities. In other words, the electricity provider was not expected to make profits to be taken out of the company – either for the state or for any other potential stakeholder. This not-for-profit principle was abandoned in the course of a reform process initiated in 1983 and, given surplus capacities, the construction of new power stations was cancelled at that time. Some operating plants were mothballed. In 2002, the national parliament passed a bill that put an end to Eskom's exemption from taxes. This has put additional pressure on the electricity supplier, because a considerable amount of money is now being transferred to the state instead of being used for build-up programmes or maintenance (Styan 2015).

What is more, ever since the first democratic elections in 1994 that brought the ANC to power, the government has been trying to transform the economy – which is, 25 years later, still dominated by the white minority. In particular, state-owned enterprises – including Eskom – have followed ambitious programmes of cooperation with non-white business partners and aimed at increasing the share of non-white employees at management level. For instance Eskom has for some years been purchasing diesel for its peak-demand power stations from black middlepersons at inflated prices, and not from BP, Chevron or Shell – which all operate refineries in South Africa (Styan 2015). Eskom, furthermore, appears to suffer from a lack of skilled managers able to negotiate sound contracts with mining companies. A confidential memorandum written in 2007 by the Canadian consultant Susan Olsen stresses that contracts signed by Eskom with suppliers specify the price and quantity of coal, but not its quality. Olsen informed Eskom that she had the impression that the firm's negotiation teams lack the necessary background and experience for successful interaction with mining companies.[4] The most demonstrative example in this regard is that Eskom pays penalties to a company called Exxaro because the Medupi Power Station is behind schedule, meaning that it cannot use the coal that is under contract with that latter company.

Problems resulting from poor management are exacerbated by corruption and personal enrichment, too. Such crimes are difficult to track, but there is certainly a close-knit relationship between energy companies and high-ranking politicians. ANC politicians own, and sit on the boards of, mining companies, most prominently President Ramaphosa himself – founder of the Shanduka Group.[5] The most important company for shale gas exploration is Shell South Africa. Twenty-eight per cent of the shares of Shell South Africa are held by the Thebe Investment Corporation, whose major shareholder is the Batho Batho Trust. Mandela once was the chairperson of that trust; Walter Sisulu, another leading ANC politician in the twentieth century, one of the funding trustees. In an interview with the *Sunday Times* (8 April 2012), Kenny Fihla, chairperson of Batho Batho at that time, acknowledged that the trust donates to the ANC.

Giving an obvious example of illicit practices, as per a report published by the *Sunday Times* (19 April 2015), Eskom has signed a contract with a mine that should not even be operating for legal reasons – because the latter does not have a valid water-use licence. The mine in question is owned by the Gupta family. The Guptas are Indian–South African businesspeople, infamous for their close relationship with former president Zuma. This case suggests that Eskom's top management – as well as the government officials who appointed them – abuse their power for the benefitting of certain businesspeople. Similar arguments have been made with regard to questionable arrangements contained in Eskom's contracts with coal mines too, especially on the quality of coal and take-or-pay stipulations (Styan 2015).

The nuclear programme has also been linked to the Guptas. In 2010, at a time of low global uranium prices, the Guptas along with Zuma's son Duduzane bought a closed uranium mine, Uranium One (later renamed Shiva Uranium). Shortly afterwards, the president started intense talks with Putin on the country's nuclear programme (*Citizen*, 17 October 2016). Finance Minister Nhlanhla Nene was supposedly fired in 2015 because he opposed this nuclear programme, arguing that it was so expensive that it would lead to the country's financial collapse (*Business Report*, 10 October 2015).

Close relations between the South African state and the private sector are not an exclusively post-apartheid phenomenon. As noted in the introductory chapter, the seminal study by Fine and Rustomjee (1996) shows how the "minerals–energy complex" also shaped South Africa economically and politically before the watershed of 1994. The complex provided crucial inputs to the manufacturing sector, and merged with manufacturing in large industrial conglomerates that benefitted from economies of scale. Generation of electricity and the chemicals industry – first of all Sasol's energy-conversion programmes – were tied to these conglomerates. Fine and Rustomjee calculated that the complex accounted for 20 to 30 per cent of GDP in the 1980s. It produced about two-thirds of all exports. The domestic financial sector served as a conveyor belt, interlinking manufacturing, mining and other sectors. The state was bound to it because of its investments in key industries such as chemicals and steel, as well as via the provision of transport infrastructure for mining and related activities.

Although there is no in-depth analysis of the post-1994 minerals–energy complex in existence, the term has nevertheless remained popular (Fine 2008; Padayachee 2010; Roberts 2007). Not all researchers agree that the minerals–energy complex is still a valid concept, though. Morris and Martin (2015) point out that there is little evidence that the post-apartheid state has been driven by mineral exporters. Mining companies have, rather, existed in constant conflict with government, in many cases resisting the government's socio-economic transformation agenda and being under an almost constant threat of nationalisation – as demanded by populist opposition parties and by some factions within the ANC itself. Industrial policy has encouraged a shift away from minerals exports. The government's economic policy concentrates on job creation through manufacturing, which is often unrelated to the conglomerates that formed the minerals–energy complex prior to 1994. Baker et al. (2015), who still believe in the usefulness of the concept, argue that the complex is somewhat compensated for by new coalitions and networks, including from the renewables sector.

McDonald argues, nonetheless, that the complex is still "vertically and horizontally integrated into a composite set of related industrial [. . .] activities" (2009: 8), with links to the service sector – finance in particular. It finds one of its most demonstrative expressions in the Energy Intensive User Group, a lobby organisation whose 31 members – including AngloGold Ashanti, Lonmin Platinum, SABMiller and Transnet – between them use approximately 44 per cent of South Africa's electricity. In a similar vein, Marquard (2006) writes that South Africa is marked by an "industrial policy elite" that advances visions incompatible with decarbonisation and reduced energy consumption. Thus, it is not surprising that the 2011 Integrated Resource Plan relies on inputs from a technical task team that consisted largely of members of government, Eskom, coal companies and energy-intensive users (Baker, Newell and Phillips 2014).

Against this backdrop, Baker (2015) argues that low-carbon initiatives – in particular, the expansion of wind energy through private procurement – are being integrated into the existing energy-intensive and high-carbon industrial frame. She shows how companies typically associated with the minerals–energy complex – for example, the aforementioned mining giant Exxaro – invest in renewable energies, often in joint ventures with multinational giants such as Tata Power from India or Vestas from Denmark. In order to meet binding criteria on socio-economic transformation in a sector dominated by white businessmen, large-scale producers from overseas – like Gestamp Wind from Spain – increasingly partner with black-owned, politically interlinked South African firms such as the aforementioned Shanduka Group – as noted, founded by President Ramaphosa. Apart from generating few benefits for local communities, this form of expanding renewables does not in any way aim at a shift in consumption patterns or rectifying the centralised nature of electricity generation and distribution. It is, moreover, tied to coal (and increasingly also natural gas) as providers of the baseload.

ENERGY-RELATED KNOW-HOW

The long-term focus on coal is reflected in energy-related know-how. Sasol has exported technologies for the conversion of coal and natural gas into chemicals, for example to Nigeria, Qatar and the United States (EIA 2015g). Considering its vast coal resources and lack of oil, India is another potential market for the South African company (Reuters, 4 June 2010). Technologies developed by Sasol could also be used to convert biomass into chemicals. The success of Sasol in general and its ability to export its technologies and start joint ventures abroad in particular depend, however, on a high global oil price.

With regard to climate change mitigation and considering the dominance of coal in South Africa's energy mix, CCS technologies are vital – at least from the viewpoint of the government. CCS means capturing carbon dioxide from large point sources such as power stations or Sasol's Secunda plant, transporting it to storage sites and depositing it where it will not enter the atmosphere – normally underground. A special report by the IPCC suggests that CCS applied to modern conventional power plants can reduce carbon dioxide emissions by 80 to 90 per cent. According to the same report, CCS has the potential to account for 10 to 55 per cent of total global carbon dioxide mitigation throughout the course of this century (Metz et al. 2005).

Against this backdrop, it is not surprising that CCS is one of the eight near-term priority programmes in South Africa's National Climate Change Response White Paper (Department of Environmental Affairs 2011). Plans to apply CCS have been jointly developed by the Department of Energy and by the South African Centre for Carbon Capture and Storage, established in 2008. The CCS Road Map, endorsed by the government in 2013, confirms the five phases of South Africa's path towards CCS. The first, completed in 2004, was about assessing the viability of the technology. During the second phase, completed six years later, potential sites for long-term storage were sourced and mapped. Applications for permits have now been made. Priority targets are depleted offshore natural gas fields in the Algoa and Zululand Basins. The third phase, which started in 2017, will lead to the construction of a pilot offshore storage plant with a capacity of 10,000 to 50,000 tonnes of carbon dioxide a year. The objective of this phase is to demonstrate that CCS can be safely undertaken in South Africa. During the fourth phase, the plant is to expand so that about 100,000 tonnes of carbon dioxide a year can be stored. It is scheduled to become a fully integrated demonstration plant by 2020. The fifth phase is about commercialising operations and reaching an annual storage capacity of one million tonnes by 2025.[6]

Introducing CCS in South Africa – or, for that matter, in any country – faces severe challenges. The aforementioned IPCC report suggests that CCS increases the cost of power plants by 20 to 90 per cent. Greenpeace has pointed out that the technology is expected to use between 10 and 40 per cent of the energy produced by the respective power stations (Rochon et al. 2008), making it unattractive to countries that run short of electricity and generally dubious concerning its supposed environmental benefits. The IPCC acknowledges that retrofitting power stations with CCS technologies will probably increase their consumption of fossil fuels by a third. As a consequence total carbon dioxide emissions may actually increase, because if a power station consumes more coal then there will be more mining and more transport, too

– both generating emissions, of course. There is also the risk of leakage, as carbon dioxide will have to be trapped for millions of years. Leaking gases would be harmful, potentially mortal, not only to the atmosphere but also to people and animals living in close proximity.

Another key issue of energy-related know-how is the nuclear build-up. Almost two decades ago, South Africa launched a nuclear programme on its own: the Pebble Bed Modular Reactor (PBMR). The PBMR company set up one of the largest research groups on nuclear power in the world. Eskom initially wanted to buy 24 units built according to PBMR design. The key advantage of the PBMR is that it uses helium as a coolant medium. Corresponding units can be built quickly. The technology is believed to be thermally more efficient and safer than water-cooled systems, such as the one used at Koeberg. However, because the technology proved difficult to handle, full operations of a demonstration plant had to be postponed from 2003 to 2009, and then eventually to 2020. In 2010, the government decided to cut financial support because the PBMR had not attracted funding from the private sector. Nor were there any private companies willing to buy nuclear plants built according to the technology developed and tested in South Africa (McKune 2010).

Provided that the nuclear programme does proceed, it is highly likely that foreign partners will bring in key technologies. In his 2015, state-of-the-nation address, President Zuma said that China, France, Russia, South Korea and the United States were potential partners for the nuclear build-up. Intergovernmental framework agreements on nuclear cooperation were signed with these countries in 2016. Several meetings between Zuma and Putin in 2015 and 2016 led to speculation that Rosatom had secured the deal even before the launch of a public tender (*Business Day*, 26 May 2016; *Citizen*, 17 October 2016; *Times Live*, 19 September 2016).

CLIMATE POLICY

South Africa's climate change mitigation strategy ought to concentrate on electricity generation and fuels synthetisation, which account for a respective 47 and 10 per cent of greenhouse gas emissions. With regard to demand-side management, the industrial sector is crucial because it is responsible for 66 per cent of that for emission-generating goods and services, following Winkler's (2007) calculations. Previous electricity shortages imposed some form of involuntary climate change mitigation. Eskom launched a demand-side management programme in 2003, which included various energy-saving measures such as replacing old light bulbs and the rolling out of solar geysers via a rebate system. In 2014, the programme was scaled down because of financial constraints (Styan 2015). Eskom bought back electricity from large-scale users up until the recent end of electricity shortages. Such measures do not, of course, constitute a strategy of climate change mitigation. Neither does South Africa's lower-than-expected GDP growth.

Winkler and Marquard (2009) divide mitigation options for South Africa into three broad categories: increasing energy efficiency; changing the energy mix; and, structural transformation of the economy. The latter category comprises numerous measures to reduce the need for energy; not only a shift away from energy-intensive industries but also, for example, urban planning that helps reduce traffic. Energy efficiency and, to a certain extent, demand-side management are suitable for short-term changes. Rapid bus systems recently introduced in certain cities and solar water-heaters are examples. The most significant steps concern energy generation, though. These will occur only in the long run. A decarbonisation of electricity generation or the introduction of a meaningful carbon tax would weigh heavily on the already overstretched financial resources of Eskom. It is uncertain that the state provider will have sufficient money for the scheduled retrofit of nine of its fossil fuel power plants. This retrofit is to reduce greenhouse gas emissions, and it constitutes not only a financial challenge; it would also require the power stations to go offline for about 120 to 150 days each, which is hardly possible considering Eskom's narrow reserve margins (Styan 2015).

In spite of these unfavourable circumstances, the promises that South Africa has made at the international level are among the most ambitious worldwide. President Zuma declared at the UN Climate Change Conference in Copenhagen that greenhouse gas emissions would be cut by 34 per cent by 2020 and by 42 per cent by 2025, as compared to a business-as-usual scenario and provided that support was provided by the Global North. The National Climate Change Response White Paper sets a benchmark for emissions, which will peak between 2020 and 2025 in a range of between 398 and 614 million tonnes of carbon dioxide equivalent. Emissions will stay on this level for up to 10 years. By 2050, they will have declined to between 212 and 428 million tonnes of carbon dioxide equivalent (Department of Environmental Affairs 2011). These objectives have been formalised in the country's INDC. Therein, South Africa declares that a transition to a low-carbon energy sector is the core element of its climate change mitigation strategy. An "inefficient fleet of ageing coal-fired power plants [will be replaced] with clean and high efficiency technology" (2015: 2), meaning more efficient coal-fired power stations and also renewable energies.

The INDC also makes clear that climate change mitigation is not a priority. It emphasises that the emerging economy faces a supply crisis – or, rather, faced one in 2015 – and stresses that South Africa is "a developing country, with overriding priorities to eliminate poverty and eradicate inequality" (2015: 2). Climate Action Tracker terms this an "inadequate" commitment (or a "highly insufficient" one in the most recent assessment). South Africa is one of the worst performers in the organisation's ranking scheme. The reason for this is that the aforementioned emissions target is equivalent to an absolute increase by 20 to 82 per cent, as compared to the level of 1990. Under current policies, greenhouse gas emissions will rather increase by 141 per cent by 2025. Climate Action Tracker criticises the fact that contracts with private generators of renewable energies have long been delayed (which is presumably a consequence of Eskom's current oversupply). The proposed carbon tax has not been implemented yet, and, during its first phase, there will most likely be exceptions for 95 per cent of emissions.[7]

What is more, the National Climate Change Response White Paper – put forward in 2011 – specifies that the aforementioned benchmark range is to be cascaded down to emissions budgets for entities and sectors within three years, with large emitting entities required to "submit mitigation plans that set out how they intend to achieve the desired emission reduction outcomes" (Department of Environmental Affairs 2011: 27). However budgets for entities and sectors have not been allocated, and no mitigation plans have been published to date. There have been discussions about a carbon tax as an alternative to company-level carbon budgets. The National Treasury introduced the idea in the early years of this century. Since then, discussion documents have been published and a carbon tax bill has been drafted. Whereas the relatively weak Department of Environmental Affairs and the National Treasury are in favour of the bill, the Department of Economic Development and the Department of Trade and Industry oppose it. Business organisations and large companies, including Eskom and Sasol, have also continued to lobby against the bill (Baker et al. 2015).

Other problematic eventualities concern clean-development mechanisms. For example, Sasol has applied for an according certificate for its 645-kilometre pipeline from Mozambican natural gas fields to Secunda. The company argues that instead of opening a new coal mine to replace the one that used to supply Secunda, the plant will now use natural gas – and thus emit less carbon dioxide. Sasol declared that the option to earn such a certificate drove its decision to build the pipeline in the first place (Earthlife 2009). Advocates of fracking in the Karoo similarly argue that unconventional resources must be tapped so as to take coal-fired power stations off the grid and avoid building new ones. Recent studies suggest that, regardless of fugitive emissions, generating electricity by burning shale gas would generate only 30 to 60 per cent of the carbon dioxide emissions that result from coal-fired power generation (Cohen and Winkler 2014). Conventional and unconventional natural gas both appear to be less environmentally damaging than coal – but one may wonder whether such a strategy means continuing on the hydrocarbon path, and thus foregoing the opportunity to switch to renewables.

NOTES

1. The presently discussed version of the Integrated Resource Plan includes planning up until 2050. It suggests that from 2030 to 2050, new nuclear power stations with a capacity of 20 GW will come online (Department of Energy 2016). Planning for such a distant future is, however, merely speculative; it does not have any immediate policy implications.

2. This scenario is based on assumptions that are likely from today's perspective: lower electricity demand than expected, competitive alternatives to nuclear energy and high costs of nuclear energy. Building nuclear power stations with a maximum output of 9.6 GW will be marginally beneficial if rather unlikely assumptions are fulfilled: high electricity demand; high costs and/or the unavailability of hydro, natural gas and renewables; and, low costs of nuclear energy.

3. The New Partnership for Africa's Development provides an overarching policy framework and vision for the continent's greater economic cooperation and integration. It was adopted by the African Union in 2001, and had been initiated by the presidents of Algeria, Nigeria, Senegal and South Africa.

4. The memo was published by the largest opposition party, the Democratic Alliance, in 2009. It is available online at: www.politicsweb.co.za/news-and-analysis/how-eskom-was-warned-of-looming-coal-supply-crisis.

5. Ramaphosa and his family trust, the Tshivhase Trust, have by now sold their shares in the Shanduka Group, but there is controversy over when this happened exactly.

6. The CCS Road Map as well as further related information is available online at: www.sacccs.org.za.

7. This information is available online at: http://climateactiontracker.org/countries/southafrica.html.

Conclusion

In the following pages, we compare the emerging economies along the five dimensions that have also structured the previous chapters. We stress that there are strong path dependencies on coal, with a certain shift towards natural gas. The tremendous potential for renewable energies that marks almost all of the emerging economies is largely untapped, in most cases due to unfavourable political and structural conditions. The same applies to opportunities for regional cooperation on energy. Another common feature is insufficient infrastructure and resulting inefficiencies (mainly blackouts and transmission losses). The focus in policy planning is, nevertheless, on expanding generation capacities, sometimes to an oversized extent. Even though the emerging economies plan to widen their renewables capacities tremendously, the reality on the ground looks different: it is mostly hydrocarbon-based power stations that are, in fact, under construction.

Such material conditions interact with non-material factors. The latter range from influential lobbies in favour of fossil fuels and nuclear energy to constrained private investment in partially liberalised markets to subsidised prices, which draw on the respective budgets of public electricity providers – making them unable to invest sufficiently in generation and transmission infrastructure. This closing chapter points out that China and South Korea have advanced visions to fundamentally transform their economies, whereas the other emerging economies stick to energy-intensive growth (albeit corresponding growth targets tend to be missed). China and South Korea are, moreover, strong on energy-related know-how. Brazil, India and South Africa have been able to advance niche technologies. Indonesia and Mexico possess no significant capabilities in this regard. Concerning climate change mitigation, we summarise that the INDCs of the emerging economies are far from sufficient for limiting global warming to 2°C at a maximum – although all emerging economies have, at least, developed sound policies relating to issues such as energy efficiency and renewables.

In the last section, we compare the relevance of different causal factors, showing that man-made material structures and political power are critical determinants of the energy and climate policies of the emerging economies – whereas naturally given material structures and energy-related know-how are rather not. We also assess how our findings may contribute to energy transition studies from a conceptual perspective. It is argued that critical junctures in the energy policies of the seven states analysed in this book seemingly means a diversification of sources, not a complete shift towards renewable energies.

MATERIAL CONDITIONS

Material conditions lay the groundwork for the energy and climate policies of the emerging economies. Electricity generation in Brazil is dominated by hydropower because of the country's enormous corresponding potential and its shortage of coal – although these are not, of course, the only reasons for this. The country's massive natural gas deposits allow for the diversification of electricity generation so as to better react to droughts, which reduce the output of hydropower stations dramatically. Exploiting this natural endowment reinforces global warming. Mexico's historical dependence on oil and gas, as well as current efforts to exploit shale gas and deepwater oil, also reflect naturally given conditions. Together with natural gas imports from the United States, policies in place cement a hydrocarbon path. Energy policy in Indonesia is shaped by the country's archipelago landscape configuration, with major challenges for the exploitation of untapped oil and gas as well as renewables – which are located on islands distant from Bali and Java, where the majority of the Indonesian people live. The proximity of Indonesia's coal reserves to the coast and relatively good transport infrastructure allow, meanwhile, for large-scale exports.

Lacking any significant domestic energy resources of its own, South Korea is one of the top energy importers in the world – in particular for coal, oil and gas. Energy imports meet 92 per cent of consumption demand (IEA 2012). The high share of nuclear power – 11 per cent, equalling a total of 23 GW (EIA 2017) – also indicates that South Korea's material conditions are different from those of the other emerging economies. In South Africa, 72 per cent of energy consumption and 94 per cent of electricity generated by the state-owned supplier Eskom derive from the country's tremendous coal reserves (EIA 2015g). Because this emerging economy does not possess any significant oil and gas resources of its own and due to international sanctions against the apartheid regime in the past, it has never made a transition from coal to oil. This has brought about an impressive path-dependent development: not only does its energy generation rely on coal, but South Africa also launched a massive coal-to-liquids programme in the early 1950s being later accompanied by a gas-to-liquids one. Today, fuel synthesised from coal and natural gas meets 86 per cent of domestic demand. The petrochemicals giant Sasol is the second-largest coal consumer, just behind Eskom (EIA 2015g).

Apart from the traditional use of biomass and waste (23 per cent), energy supply in India is dominated by coal and peat (both 46 per cent) as well as by oil (22 per cent; all figures from IEA data). Whereas coal resources are abundant, being the fifth-largest in the world and making India the third-largest producer, the emerging economy runs short on oil, leading to a high dependence on corresponding imports. More than 90 per cent of India's coal is produced by open cast mining (IEA 2015a), which is cheaper than deep mining and thus partly explains why coal is so important to the country. China's energy mix is also dominated by coal. The People's Republic will remain the world's largest coal consumer for the foreseeable future, accounting for half of global consumption today and an estimated share of slightly more than 40 per cent by 2040. Domestic resources amount to 13 per cent of world reserves (IEA 2016a). Oil – the second-most important energy source – is increasingly being imported because of depleting domestic reserves. For electricity generation, hydropower plays a significant role, reaching a share of 22 per cent – reflecting natural conditions, meaning the significant availability of rivers suitable for building dams and associated power stations.[1]

Even though all of the emerging economies – with the exception of South Korea – benefit from some form of resource endowment, many face major challenges in exploiting their own domestic reserves. Political problems aside (more on these later), there are also unfavourable geographical conditions such as long distances from sites of resource extraction to centres of population and economic activity, rather poor quality to resources as well as other key obstacles to exploiting them. These include geological barriers (for geothermal and hydrocarbon resources), unfavourable landscape configurations (for solar and wind energy), sedimentation in rivers (for hydropower) and insufficient infrastructure for transmission/transport. It appears that coal and hydropower are relatively easy to exploit, whereas oil and gas more challengingly so because of high capital costs – as especially Indonesia and Mexico demonstrate, where potentially large deposits remain unexplored (*Oil and Gas Journal*, 8 May 2017; PwC 2014).

This constellation accounts for a dominance of coal-fired power stations in most of our cases, apparently leading to strong path dependencies – plants cannot be taken offline overnight, at least not without suffering heavy monetary losses (and without meeting with strong resistance from influential economic and political actors in favour of fossil fuels). There are 132 active coal-fired power stations in India, some reaching a capacity of 2 GW or more. Several large-scale thermal plants have recently become operational in Indonesia and South Africa, or are being constructed there. They will still emit carbon dioxide in the second half of this century. However, most of South Africa's large 12 coal-fired power stations (output of 1.2 to 4 GW each) will be decommissioned in the 2020s and 2030s, indicating a critical juncture in its energy policy. South Korea's six giant coal-fired power stations, conversely, are unlikely to go offline anytime soon. Their first units were built from 1983 onwards. The others are of recent design. Building up its infrastructure for LNG-based electricity generation, South Korea appears to have set out on another development path of hydrocarbon dependence. Brazil and Mexico have undertaken major efforts to exploit their domestic oil and gas reserves, instead of shifting to renewables. The Chinese government, meanwhile, has been struggling not only to increase the efficiency of its vast fleet of thermal plants but also to close the least efficient of the thousands of coal mines currently operating in the People's Republic today.

Regarding renewable energies, there is a tremendous gap between potential and actual exploitation. In Mexico, the enormous potential of solar and wind energy is scarcely being tapped into. The Central American country arguably possesses the fifth-largest geothermal potential globally. Average solar irradiation reaches some of the highest values worldwide, and the Isthmus of Tehuantepec is one of the windiest places in the world (EIA 2016b; IEA 2016c). The gap between potential and actual use is also evident for solar energy in Brazil, which could play a major role for mini-grids in the sparsely populated north and for providing electricity for air conditioning in urban areas (Ramos Martins et al. 2008b). Still, solar energy contributes only 0.01 per cent to Brazil's electricity generation. Although absolute generation capacities are impressive when compared to other countries, China hardly makes use of opportunities for solar and wind energy. Biomass could be exploited to a much larger extent, too. In terms of hydropower, the People's Republic has tapped almost 300 GW out of a technical potential of 400 to 700 GW (IRENA 2014; REN21 2015). Indonesia possesses a considerable potential for hydro, solar and wind energy, but it uses little of any of these. The same applies to the tremendous opportunities for geothermal electricity generation: using such energy instead of fossil fuels in Indonesia would equate to the largest carbon dioxide emissions reduction project in the world (Sukarna 2012).

In India, three-quarters of the total hydropower potential are untapped – equivalent to 149 GW (Planning Commission 2013). Only a minuscule share of the potential of biomass, solar and wind energy is currently being used. As in the case of China, much of this potential is located in remote areas and it would be very expensive to connect them to the national electricity grid. In South Africa, the pulp industry and also sugar cane processing offer opportunities for bioenergy. More than 80 per cent of the country has sufficient natural conditions for economically viable wind farms, while high solar irradiation offers up tremendous potential (Fluri and von Backström 2009; Knorr et al. 2016) – which could allow for local-grid electrification in rural areas. Yet, at present, non-hydro renewables account for less than 1 per cent of total energy consumption (EIA 2015g). The case of South Korea appears to be different. Although recent assessments by the government conclude that the country has considerable technical potential for renewables, about 70 per cent of South Korea is mountainous. Installing solar and wind power facilities there would be costly. In the flat parts of the country, population density is high – leaving little space for power plants being built there. Offshore wind power is hampered because of concerns over fishing and military installations (Duffield 2014).

Biofuels are hardly relevant for most of these emerging economies. They appear to be neglected in South Africa. South Korea's KETEP mentions them as a topic among many others for research related to green growth. India's Integrated Energy Policy of 2006 refers to biofuels, but without going into details. In China, biofuels – largely based on ethanol – have not gained considerable relevance yet. For Mexico, biofuels play a marginal role (0.8 per cent of the energy used in the transport sector) and will remain more expensive than conventional fuel for the foreseeable future – even in spite of subsidies for conventional diesel and petrol being cut (IRENA 2015a). Biofuels do matter significantly in Brazil, where car engines burn petrol blended with 20 to 25 per cent of ethanol or pure ethanol. This is the outcome of several decades of public subsidies: Brazil's programme ProAlcool was launched in response to the oil crises of the 1970s. Indonesia introduced mandatory biodiesel blending rates of 10 per cent in 2013, 15 per cent in 2015 and 20 per cent in 2016.

Another feature that applies to most of the countries analysed in this book is transmission infrastructure in urgent need of repair. Forty-seven per cent of Mexico's sparse transmission lines are more than 20 years old by now (Alpizar and Rodríguez 2016). Losses are equivalent to 14 per cent of electricity generation, mostly due to non-payment and theft though (IEA 2016c). In Brazil, distribution and transmission losses amount to 16 to 17 per cent (World Bank 2014). Long distances and the often disparate locations of hydropower stations are major challenges. Outdated transmission lines account for tremendous losses in India. There are also considerable commercial losses due to insufficient billing, poor metering and to theft. Counted as one, these losses amounted to 23 per cent in 2014 (Swain 2016). A lack of LNG terminals and pipelines hampers the use of natural gas (instead of traditional biomass and coal). Investments in South Africa's transmission grid are insufficient. Even worse, municipal authorities are obliged to pay for the maintenance and expansion of electricity distribution infrastructure – but usually fail to do so (Styan 2015). China's power grid does not suffer from these kinds of problems, but it does need to be expanded considerably to allow for a better use of solar and wind energy. In Indonesia, the archipelago landscape configuration causes enormous challenges for the power grid; so do institutional shortcomings. Due to a lack of pipelines, vast amounts of natural gas are flared instead of being consumed locally (McKinsey 2014a).

Opportunities for regional cooperation, based on natural conditions, exist for most of the emerging economies – Mexico and South Korea are exceptions in this regard. Yet, these possibilities are largely foregone. Indonesia participates in the TAGP, a project by Indonesia, Malaysia, Myanmar, Singapore, Thailand and Vietnam that aims at interlinking natural gas reserves in South East Asia. Presently, Indonesia does not have any cross-border oil or gas pipelines though. The TAGP is accompanied by the APG, which is to connect the electricity grids of Cambodia, Indonesia, Laos, Malaysia, the Philippines, Thailand and Vietnam. The APG currently remains rather vision than reality (Ahmed et al. 2017). Tapping energy resources in its wider neighbourhood has been a political vision for South Africa ever since the country's first democratic elections in 1994. Angola, Mozambique and Tanzania are rich in natural gas. Congo-Kinshasa possesses tremendous hydropower potential. Yet, the only successful larger project ever witnessed there dates back to the colonial era, when a power station at the Cahora Bassa Dam in Mozambique was connected to South Africa. Recent projects have not led to significant imports of electricity, and it appears that, since the 2008 energy crisis, Eskom and the government have focussed on domestic solutions, as these are more likely to deliver reliable results (Scholvin 2015).

Brazil imports vast amounts of natural gas from Bolivia, and operates the world's second-largest hydropower station (in collaboration with Paraguay). More ambitious projects have either failed – for example, a transcontinental natural gas pipeline – or are in an early stage of development, such as large-scale hydropower schemes with Argentina and Bolivia (COSIPLAN 2016). The Brazilian case demonstrates the risks of depending on energy imports from neighbouring countries: prices for natural gas from Bolivia have increased considerably over time, and nationalisation policies have threatened Brazilian investments there. India's efforts to acquire electricity generated by hydropower stations in its neighbouring countries as well as natural gas (via pipelines from Iran and Myanmar) have not led to any significant positive results. After being stuck in limbo for years, several projects do appear to have gained momentum again under the Modi government however. Without formal backing from the government, a consortium of Indian companies has advanced plans to build an underwater pipeline to Iran – thus bypassing Pakistan (*First Post*, 6 September 2017).

What is more, Indian firms have expanded globally in order to acquire equity in upstream oil, gas and – to a lesser extent – coal projects. This applies even more to China, whose state-owned enterprises have outcompeted Indian rivals in many instances. Regardless of Chinese relative success, neither its nor India's efforts to increase energy security via acquiring assets abroad have been a major success. From a purely economic perspective, this strategy has proven costly and ineffective: the share of resources extracted abroad by Chinese and Indian companies respectively has remained small in comparison to both domestic consumption and conventional imports (IEA 2014a; Planning Commission 2012). The same problems apply to the efforts of South Korean state-owned firms to acquire oil and gas blocks abroad. These additionally suffer from increasing debt, as a result of credit intake for overseas acquisitions (EIA 2017; Jeon 2011).

ENERGY DEMAND AND EXPANSION PLANS

As noted in the introductory chapter, the emerging economies are already major energy consumers. Their energy consumption will, furthermore, increase considerably in the coming decades, which makes them critical players for climate change mitigation. Brazil has more than doubled its electricity generation during the last 20 years. Prior to the country's economic crisis, it was argued that Brazil would need 6 GW of new capacity each year until 2035 (IEA 2013). Following pre-crisis plans, 156 GW of hydropower, 21 GW of nuclear power, 20 GW of non-hydro renewable energies, 7 GW of natural gas and 6 GW of coal are to go online by 2035 (Ministério de Minas e Energia 2007). Plans written during the midst of the crisis suggest more modest expansion. Indonesia's primary energy consumption grew by 43 per cent between 2003 and 2013 (EIA 2015e). Energy demand is likely to increase further, as the government aims to transform the country into one of the top-10 economies globally. By 2025, Indonesia may need an electricity generation capacity of 430 to 550 GW – meaning 10 times its current capacity (Ibrahim 2014). A fast-track expansion programme from 2006 aims at building new coal-fired capacities of almost 21 GW until 2022. On top of this programme, the latest electricity-supply plan calls for 35 GW to be added by 2019 (56 per cent by coal, 36 per cent by natural gas).

According to the 2011 version of South Africa's Integrated Resource Plan, electricity supply will henceforth expand and diversify. Forty-two per cent of the additional electricity – that is, 17.8 GW – shall be generated by solar and wind energy by 2030. Nuclear energy is to account for 23 per cent of the new capacities, or 9.6 GW; coal for another 15 per cent, or 6.3 GW. Hydropower and natural gas play scarcely important roles (Department of Energy 2011). This planning is, however, outdated, because the South African economy has since grown more slowly than expected. The nuclear build-up will probably not happen at all, and the tremendous expansion of renewables appears difficult to achieve considering the slow progress in this regard during the last couple of years. As said, natural gas – imported from certain neighbouring countries – constitutes an interesting alternative. The non-ratified 2013 version of the Integrated Resource Plan accordingly specifies that natural gas will contribute the largest share to yet-to-be-built capacities (12.7 GW), followed by wind (6.4 GW), coal (5.3 GW) and solar energy (4.7 GW) (Department of Energy 2013).

Although South Korea is further developed than the other emerging economies, its energy consumption is still increasing – in contrast to corresponding dynamics in other OECD countries. South Korea's first national Energy Master Plan, which covered the years from 2008 to 2030, projected an increase of primary energy demand by nearly 50 per cent. It stated that the share of fossil fuels would decline to 75 per cent by 2030 (or to 61 per cent in a more aggressive scenario). Natural gas was projected to gain importance – but not coal and oil – and the share of renewables should reach 11 per cent. The second Energy Master Plan did not propose significant changes to this outlook. According to this plan, the share of renewable energies in energy consumption was to grow fastest (4.4 per cent a year), followed by nuclear power (3.3 per cent) and natural gas (1.9 per cent) (Ministry of Trade, Industry and Energy 2014). The more recent 8th Basic Plan for Electricity Supply and Demand, conversely, foresees an increase in the share of renewables to 20 per cent of electricity supply by 2030 as well as a reduction of coal to 36 per cent and nuclear power to 24 per cent.

India is the emerging economy with the fastest growth in electricity demand. The Planning Commission (2013) expects energy demand there to increase by 5.4 to 5.7 per cent a year until 2022 and to have quadrupled by 2035, as compared to the 2005 level. The successor institution of the Planning Commission, the Policy Commission, predicts even higher annual growth rates of 5.7 to 7.4 per cent (NITI Aaayog 2015b). According to the Draft National Energy Policy, coal-based power generation is to increase considerably – reaching a total of 330 or even 441 GW by 2040. Coal mining will double (NITI Aayog 2017). However, six projects related to new coal-fired power plants were cancelled or shelved in 2015, because current electricity demand did not justify their construction anymore. Very optimistically, the draft policy also sets a high target for renewables: 597 to 710 GW by 2040. Already by 2030, 30 new nuclear reactors are to become operational – bringing total nuclear capacity to 63 GW. Considering past delays, current legal disputes and increasing public opposition, this objective is highly unrealistic.

China's energy consumption and generation reach nearly twice the respective amounts of all our other cases combined. The People's Republic has contributed about half of the worldwide growth in energy consumption and generation for a decade now. The Energy Development Strategic Action Plan now sets a cap for coal consumption at 4.2 billion tonnes, meaning that annual production must decline by 2.9 per cent and the share of coal in the energy mix needs to drop to 62 per cent. The 13th Five-Year Plan reiterates the cap on coal, and sets a target of 15 per cent for the share of renewables in the energy mix by 2020. This equals a total contribution by hydropower of 350 GW, by solar power of 100 GW and by wind power of 200 GW. Nuclear power shall also expand rapidly: to a total generation potential of 58 GW.

Mexico's electricity demand increased by an annual average of 2.7 per cent between 2005 and 2015. The IEA (2016c) predicts that such demand will increase by 85 per cent until 2040. A study by McKinsey (2014) suggests that energy demand could be 79 per cent higher in 2024 than it was in 2010. Following the government's plans, coal is to dominate electricity generation by 2030, with a share of 58 per cent, followed by hydropower (11 per cent), wind energy (11 per cent) and nuclear energy (9 per cent). Oil will become less relevant than it presently is (SENER 2016). Legislation specifies that 35 per cent of energy generated by 2024 is to come from renewable sources (Cámara de Diputados 2016). Power stations with a joint maximum output of 57 GW are to be built in the course of the next 15 years. Solar and wind energy are to contribute a respective 12 and 21 per cent of these capacities (SENER 2016). There is some uncertainty about the validity of these projections because the recent liberalisation of Mexico's energy sector will change investment dynamics, with natural gas and/or non-hydro renewables most likely benefitting.

Beyond mere energy generation, China and South Korea appear to be the only emerging economies seeking to fundamentally change their economies. China's paradigm of a "new normal" not only means a successive reduction of the GDP growth rate but even more so a shift to less energy-intensive products and services. It favours domestic consumption over exporting manufactured goods. This includes a focus on high-quality growth, meaning innovation, low-carbon technologies and the reduction of socio-economic inequality. These objectives are reflected in recent five-year plans and other official documents, which specify how the restructuring of the Chinese economy is to be achieved. Special programmes have been implemented to increase energy efficiency. Waste-to-heat utilisation and the closure of small, inefficient factories are other examples in this regard. The rationale behind

South Korea's own "low carbon, green growth" vision is that the country's development trajectory is not sustainable and ought to shift to one of quality-oriented growth – with a focus on energy independence and sustainability achieved through an increasing use of renewable energies (Choi 2015). As we explain below, these objectives become particularly apparent in research and development.

Political convictions in South Africa are different: in numerous policy documents socio-economic development is associated with GDP growth (Department of Finance 1996; Economic Development Department 2010; Presidency 2005), which is linked in turn to increasing energy demand. Prior to the UN Climate Summit in Paris, India's government emphasised that its focus is economic growth and poverty reduction; so did South Africa in its INDC. India has announced a doubling of coal production until 2019. Referring to the objective of making the country one of the top-10 economies globally, Indonesia's National Energy Council predicts per capita electricity consumption to increase from 650 kWh in 2010 to about 7,000 kWh in 2050 (Ibrahim 2014). Brazil, which does have comparative strengths in green growth sectors such as biofuels, also concentrates on economic growth, seeking to develop its pre-salt oil and gas resources instead of working towards a structural decarbonisation of its economy. The new president has called all climate change mitigation efforts into question, meanwhile.

Furthermore, one ought to be careful with confusing terminology. Mexico's government claims that it pursues green growth objectives. This does not, however, translate into a fundamental change of the economy so as to decarbonise energy generation and shift to less energy-intensive production. Green growth in Mexico means minor adjustments such as less natural gas flaring and better control of methane leaking in a still highly carbon-dependent economy. President Peña Nieto, whose mandate had just ended when the manuscript of this book was being finalised, emphasised his desire for his country to make use of its natural resource abundance – mainly oil and gas – as a driver of economic development.

POLITICO-ECONOMIC CONSTELLATIONS

The energy and climate policies of the emerging economies are shaped by a combination of factors, and even though we have emphasised material conditions throughout this book we nonetheless do not seek to provide geo-determinist explanations for these. Rather, we are convinced that material conditions interact with economic, political and social variables, forming combined causal factors that account for the outcomes witnessed in energy and climate policies. Focussing on the interaction of material and non-material conditions means recognising that there are various options available to governments and influential societal forces in shaping energy and climate policies (at least at critical junctures, which allow a shift from one path or regime to another), thereby partly overcoming constraints imposed by man-made and naturally given material structures.

In Mexico, nearly 16 GW of ageing capacity – out of a total of 65 GW – is to be decommissioned by 2030. About 10 GW of this is conventional thermal. The Mexican government appears to have chosen to stick to hydrocarbons, building power stations fired by natural gas instead of investing more in renewable energies. Most of South Africa's coal-fired power stations will, as noted, be decommissioned in the 2020s and 2030s. Because of climate change, and considering that coal mining does not bring returns as quickly as required, it would make much more sense for South Africa to shift to other sources of energy. Natural gas could be imported from Angola and Mozambique. As noted, regional cooperation on hydropower is another extremely viable option; so is the exploitation of the domestic potential for biomass, solar and wind energy. Similar opportunities exist to some extent in China, India and Indonesia because of the planned enormous expansion of these countries' electricity-generation capacities, although critical junctures are exceptional opportunities. In this sense, Brazil appears to be on a cemented non-renewables track. In South Korea, the Moon government has promised to shut down several coal-fired and nuclear power stations; it remains to be seen, however, to what extent these promises will actually become reality.

Another example of political decisions being critical for energy and climate policies is subsidies. This does not appear to be a major issue for Brazil, although the country has recently suffered from violent protests against rising diesel and petrol prices (*O Globo*, 31 May 2018). In India, coal prices are set by public companies – first and foremost Coal India – giving preference to industrial consumption. This system has generated an implicit subsidy for local consumption that amounts to USD 2 to 10 billion a year (IEA 2015a), putting other energy sources at a disadvantage. Retail prices for electricity in South Korea are subsidised to protect poor customers and even more so agriculture and the industrial sector, for which prices are set at 33 to 50 per cent of the average one (Chung 2014; IEA 2012). In Indonesia, energy subsidies amounted to almost 11 per cent of state budget expenditure and 2.5 per cent of GDP in 2013 (IEA 2015b). Efforts to scrap these subsidies have been undertaken, usually as a result of budget constraints – and not in order to decrease energy consumption.

China has recently increased tariffs for urban middle-income households, but cheap electricity continues to be seen as a tool to maintain social stability. Hence, the production of fossil fuels is subsidised by budgetary support for state-owned energy companies, freight and tax discounts, as well as government support for corresponding research – estimated to reach slightly more than USD 3 billion a year (Bath et al. 2015). In South Africa, electricity tariffs have risen considerably during the last 10 years. Electricity theft (through unauthorised connections to the grid) and non-payment are widely tolerated. Electricity and fuel used to be heavily subsidised in Mexico, particularly for agriculture and private households. Electricity subsidies amounted to 0.6 per cent of public spending in 2015 (IEA 2016d). The government had previously taken steps to cut back on all energy subsidies, which had, occasionally, provoked violent protests.

A similarity across some of the cases analysed in this book is that energy subsidies – or, rather, the necessity for someone to ultimately carry the burden of these – have dramatically reduced the capacity to invest in generation and transmission infrastructure. Electricity prices in India are set by the federal states, and since they are afraid of social unrest then tariffs do not usually reflect what power providers themselves must invest. The same logic – with an additional focus on the competitiveness of energy-intensive industries – applies to Indonesia and South Africa, where electricity prices are set by the national government. This problematic situation has been reinforced in India by rail freight – including coal – cross-subsidising commuter transport, and thus being overly expensive. In South Africa, municipalities should maintain and upgrade electricity distribution systems – but lack the funds to do so.

The dominant role of state-owned energy providers in many of our cases hampers private investment. Beyond Eskom's power-purchase agreements with private firms, no steps towards liberalising South Africa's electricity market have been taken. In spite of recent decisions taken to liberalise South Korea's electricity and natural gas markets, state-owned KOGAS remains responsible for 95 per cent of the country's natural gas imports and sets end-user prices – ones high by OECD standards. Electricity generation, transmission and distribution are controlled by state-owned KEPCO, meanwhile, and whereas heavy price subsidies have been granted for agriculture, the industrial sector and poor customers, other electricity users pay prices above the OECD average. The electricity sector in Brazil, conversely, is fairly liberalised. All sources of electricity have to compete against each other in bidding processes. Distributors and large-scale consumers are obliged to enter into long-term purchase agreements, which are made through auctions. Local content obligations for the oil and gas sector, including the obligatory participation of Petrobras in new pre-salt exploration projects, have recently been eased so as to guarantee investment in times of domestic economic crisis. Private and foreign investment in China's electricity market was first permitted in 1997, but corresponding dynamics never took off because of a lack of confidence in the country's continued growth, misgivings about power-purchase agreements with local counterparts and a rapidly declining domestic need for capital, management skills and technologies from abroad.

Mexico has liberalised its energy sector (without privatising the oil and gas giant PEMEX, or the electricity provider CFE). Through private investment in oil and gas extraction and processing, in electricity generation as well as energy in transmission networks, end-user prices are to decline. There is, however, little public support for this liberalisation, and the CFE and PEMEX remain privileged players in the electricity and hydrocarbon markets respectively. In Indonesia, the Constitutional Court overthrew a law that would have contributed to liberalising the electricity market in 2002. Two years later, its judges argued that market pricing would prevent the government from ensuring energy supply and make Indonesia vulnerable to price manipulation by foreign powers (Mochtar 2015). They reasoned, in a verdict from 2012, that all natural resources should be "controlled by the state to be exploited to the greatest benefit of the people" (quoted in Stuart 2015: 19).

Private companies have been able to invest in India's upstream oil and gas sector since the late 1990s, but they have hardly done so because the down- and midstream sectors remain largely state-controlled. The government also determined prices for refined products until the beginning of this decade, usually below market values (EIA 2016a). Natural gas cannot be sold on a purely commercial basis. It is allocated to different consumers, with priority given to fertiliser production and power generation. Since 2003, private companies may freely generate, transmit and sell electricity to end users, but federal states and their state-owned grid operators have been trying to keep private generators out by levying high transmission charges or outright denial of access (Singh 2016). Coal mining has not been liberalised, as foreign companies are only allowed to produce the commodity for their own needs – not for the general market. Even in these cases, Coal India identifies the blocks for private participation, and investors have to transfer 26 per cent of mining revenue to affected communities (Ebinger 2016).

Most of the emerging economies use feed-in tariffs, clean energy certificates, minimum shares of renewables (for generators and end users) and other similar measures to support the expansion of environmentally friendly energy. Prices for electricity from renewable sources are subsidised in Indonesia. The development of solar and wind power is supported by tax incentives, mandatory feed-in regulations and preferential tariffs in India, so as to make these sources of electricity competitive by 2022. Large utilities are obliged to acquire certificates for renewable energy generation, the so-called renewable energy certificates (RECs), covering a minimum of their respective total generation. These measures have not, however, been overly effective so far (Shrimali, Tirumalachetty and Nelson 2012). Brazil had a feed-in tariff for biomass, small-scale hydropower and wind from 2002 to 2011. It allowed for 2.6 GW to be deployed (Luomo 2014). Presently, renewable energies are supported by tax exemptions and low-cost financing. All this works somewhat differently in China, where the central government pushes the provincial ones to achieve set targets on renewables' deployment through a variety of means, including preferential feed-in tariffs.

Feed-in tariffs, granted between 2002 and 2012, have allowed for the rapid development of renewable capacity in South Korea, which has, though, occurred at a very low absolute level. Afterwards, South Korea shifted to a different system, obliging power generators to reach a share of 10 per cent of renewables in their respective generation by 2022 – starting with 2 per cent in 2012, and rising by 0.5 per cent each year thereafter. So far, annual targets have not been met (Global Green Growth Institute 2015; Yoon and Sim 2015). In South Africa, Eskom signs long-term purchase agreements with private generators of renewable energies. These have to compete against one another in auctions, and will soon generate slightly more than 6.3 GW of almost exclusively solar and wind power (Eberhard and Naude 2016). Mexico plans to introduce clean energy certificates, the sale of which shall be invested in renewable energy projects. The emerging economies have also taken steps to support biofuels, usually by making specific blending rates mandatory; but, as observed above, the impact of these measures remains marginal except for Brazil, which has supported biofuels since the 1970s.

As all this indicates, the implementation of policies is a major problem for the emerging economies. One of our key findings is that several of the emerging economies do not possess adequate institutional arrangements for sound energy and climate policies. South Korea is the exception here. Its Ministry of Trade, Industry and Energy is the sole body responsible for energy policy, including climate change mitigation, energy saving, price control, resource development and the general supervision of industry. Two commissions – the Presidential Committee on Green Growth and the National Energy Committee – have issue-specific mandates. Private sector representatives are neatly incorporated into these structures. Brazil does not suffer from institutional overlap and unclear competences, but from major corruption. Given the central role herein of Petrobras, the resulting crisis has brought activities in the upstream oil and gas sector to a standstill recently. Somewhat similarly, South Africa's failure to advance a better energy policy is not due to formal deficiencies in the political apparatus. It rather results from corruption and close personal ties between businesspeople and politicians, as we explain at the end of this section.

In India, meanwhile, five ministries together with the Energy Coordination Committee are directly involved in energy policy. The federal states hold decisive competences for certain critical issues, such as natural gas infrastructure. Decentralisation in Indonesia has bestowed considerable responsibilities on local governments. They are critical for the expansion of geothermal energy, among other core issues. However, many of these public authorities lack the capacities needed to draft and implement appropriate policies. They also run short of capital for corresponding investments, including subsidies and guarantees for the private companies that would carry out geothermal exploration. Furthermore, local governments in Indonesia are driven by the short-term motivation of securing political support – for example, through public spending on education and health – instead of by long-term visions regarding energy policy and climate change mitigation (Winters and Cawley 2015). Applying a wider definition of institutions, Mexico's Energy Reform does not guarantee that the incoming government will continue with the liberalisation agenda. Without changing legislation, it may restrict foreign investment and reinforce the role of state-owned companies – especially in the upstream oil and gas sector.

The mapping out of China's energy policy and the clarity of its corresponding course suffered in the past from the fragmentation of state-owned enterprises, the relative policymaking autonomy of provinces and energy companies, considerable administrative overlap and from insufficient overall coordination. These problems have been addressed, but by no means fully fixed. Provinces still seek to avoid the closure of inefficient factories and mines and to prevent the reduction of coal output within their own geographical boundaries (Kong 2011; Williams 2014). Whereas government agencies mandated to coordinate and implement energy policy run short of the financial and human resources needed for fulfilling these tasks, state-owned energy firms are actively involved in the drafting of energy laws and have occupied key positions within supervisory bodies. Close personal links exist between the executive boards of energy companies and the national government. Most importantly, several members of the central committee of the Communist Party began their career in the oil industry indeed. It appears that oil firms delay the central government's energy-intensity targets and resist the addition of renewable energy to the grid (Williams 2014). In the last few years these companies have become largely independent from political oversight, pursuing their own foreign economic policies, initiating deals and making overseas investments at times and in places when/where this was discouraged by the political leadership itself (Baumgartner and Godehardt 2012; Meidan 2016; Müller-Kraenner 2008).

Vested interests – sometimes involving corruption – also play a considerable role in the other countries, too. South Korea's nuclear lobby is very influential. Brazil's economic and political crisis is the outcome of a level of corruption that encompasses leading politicians and, as a key protagonist, Petrobras, as mentioned. In India, corruption in granting mining licenses is a core legacy of the Singh government. Coal India remains highly influential in politics, in particular because it is the second-largest employer in the country – with a workforce of 370,000 people. The directors general of Mexico's oil giant PEMEX are short-term presidential appointees, indicating close ties between energy firms and government. In Indonesia, local politicians and black market traders collaborate in illegally selling subsidised diesel. They are not only opposed to a reduction in subsidies for it, but also work against projects that support alternative sources of electricity for rural Indonesia (Winters and Cawvey 2015).

The expansion of nuclear energy in South Africa is unnecessary to meet domestic electricity demand and would, if carried out, come along with serious financial repercussions; neither the state nor the state-owned electricity provider have sufficient capacities to pay for the planned 9.6 GW of additional nuclear power. Yet, it appears that former president Zuma pushed the nuclear build-up to create economic opportunities for members of his own family and for favoured businesspeople (*Citizen*, 17 October 2016). Coal contracts signed by Eskom are believed to reflect such forms of corruption and personal enrichment, too (*Sunday Times*, 19 April 2015). More generally speaking, and moving away from the realm of illegal business activities, South Africa is marked by what researchers have called an "industrial policy elite" or a "minerals–energy complex" (Fine and Rustomjee 1996; Marquard 2006) – although the latter concept has become somewhat outdated by now. This means that politically dominant interest groups and networks make sure that fossil fuels – in particular coal – and energy-intensive industries remain at the core of the country's economy, to the disadvantage of renewable energies.

ENERGY-RELATED KNOW-HOW

There are considerable differences between the emerging economies with regard to energy-related know-how. These are largely due to the fact that some have invested heavily in corresponding research and development, whereas others have not. On the one side, China has become the world leader in production of solar and wind energy components. This is a result of its public investment in corresponding research and design, subsidies for domestic companies, local content obligations for foreign investors (in particular with regard to sharing technologies) as well as the existence of significant exports and a large home market. In 2012, Chinese companies accounted for 71 per cent of global production of photovoltaic cells – albeit with increasing overcapacities (Puttaswamy and Sahil Ali 2015; REN21 2015). By 2011, the People's Republic had become home to 80 producers of wind installations, 8 of which were among the top 15 worldwide. Nuclear fusion, arguably ready for commercial use in China in 50 to 60 years' time (*South China Morning Post*, 17 July 2017), is probably the most important area of energy-related know-how for the People's Republic from a long-term perspective.

South Korea has diverted 25 per cent of its expenditure for research and development to green sectors, in order to advance its vision of low carbon, green growth – therein applying a broad definition of green sectors that also covers CCS and nuclear energy. This is particularly important because South Korea is among the OECD countries with the highest share of investment in research and development, spending 4.2 per cent of its GDP thereon. A road map and strategic plan for research and development on energy-related sectors, items and technologies have been developed so as to guide corresponding investment decisions with year-by-year plans. The objective is to make South Korea a global leader in green technologies, and to capture a significant portion of the markets for green technologies. A strategy for green-technology commercialisation narrows the approach somewhat down, focussing on those in seven specific sectors: CCS, fuel cells, hybrid cars, LED lighting, light-water reactors, smart-grid systems and solar cells. Already today, South Korea is a world leader in nuclear energy technology and it aims to export (mostly to the Global South) 80 nuclear reactors with full intellectual property rights by 2030 (World Nuclear Association 2017b).

Standing in the middle on a scale of innovativeness, Brazil possesses considerable strength in research on ethanol (and applications thereof) and on offshore oil extraction. It obliges private energy firms to spend 1 per cent of their net income on research activities. Market opportunities for hydropower have triggered major foreign investment, with overseas companies now carrying out research and development in Brazil. Given their domestic experience, Brazilian engineering companies have considerable prospects to become involved in hydropower projects abroad (*China Daily*, 6 May 2017). South Africa has certain, albeit somewhat debatable, strengths in niche technologies: coal-to-liquids conversion (because of Sasol's several decades of experience) and small-scale nuclear power stations (namely, the failed PBMR programme). An impressive 22 per cent of spending on energy-related research in India is destined for nuclear energy, although total expenditure is low by international standards. India manages the entire nuclear fuel cycle. Present research is focussed on using thorium as a fuel, and India is actually the world leader when it comes to commercialising this technology. It has been argued that the rather small Indian thorium-based reactors are suitable for export to countries with a low total electricity demand (*Guardian*, 1 November 2011). Presently, only one prototype has been constructed though. Commercial use is expected by around 2030.

At the other end of the spectrum, Indonesia and Mexico have not made significant investments in research and development related to energy generation. They depend on foreign investment or, as far as feasible, domestic companies purchasing foreign technologies. In the case of Indonesia, considerable opportunities regarding geothermal power and related research are foregone. Finding ways to make use of LNG for boats and ferries, criss-crossing Indonesia's territorial waters and connecting the country's practically countless islands, could also constitute an area for innovation, but we are not aware of any corresponding efforts currently being undertaken. Mexico's institutions that carry out research on energy-related topics – the five Centres for Energy Innovation – suffer from having only a very small budget, of USD 160 million (Bonafé, Beltán and Neira 2016).

An issue of relevance to all of the emerging economies – with the exception of Brazil – is CCS. The Mexican government considers plants equipped with CCS technology to be clean energy and will, most likely, seek to apply it widely. CCS has been considered part of Indonesia's national energy policy, and pilot projects do exist. No concrete steps towards implementation have been taken, however. Companies in China have received strong government support for research on CCS since the inception of the 12th Five-Year Plan. A first pilot project, a combined heat and power plant, started operations in 2011. It is expected that China will have commercial-scale CCS demonstration projects by around 2020 (ENGO Network on CCS 2015). In South Korea, CCS has benefitted from generous public financing since 2009. Demonstration projects have been completed. Storage sites – including some offshore – have been explored, and combustion technology has been developed. CCS is one of the eight near-term priority programmes described in South Africa's National Climate Change Response White Paper (Department of Environmental Affairs 2011). Sites suitable for long-term storage have been identified, and a pilot offshore storage plant is now being constructed. By 2025, commercial operation is to commence. Interest in CCS among private and state-owned firms in India is, meanwhile, on the decline, because of both regulatory hurdles and public opposition to underground storage (Global CCS Institute 2013).

CLIMATE POLICY

All of the emerging economies have published policy plans on climate change mitigation, especially with regard to expanding renewables – thereby increasing energy efficiency and reducing greenhouse gas emissions. South Korea's Target Management Scheme, in force between 2011 and 2015, and the subsequently implemented Emissions Trading Scheme envisage a significant reduction of consumption among the country's most energy-intensive companies. Several other measures – energy-efficient construction and mass transport, for example – have been introduced. India's 12th Five-Year Plan states that the efficiency of existing thermal power plants ought to increase. Half of the new thermal power stations from that plan and its predecessor are to be supercritical. Since the early years of this century, the government can prescribe energy-consumption norms and standards for commercial buildings and energy-intensive industries. A plan regarding low-carbon strategies for energy-intensive appliances and industrial sectors has been developed.

China has implemented more and more encompassing reforms than any other country dealt with in this book. It has also been more successful in reaching its own official targets for energy saving and emissions reduction. The People's Republic has been one of the first states globally to use internationally recommended market mechanisms to support a shift in energy generation. The 12th Five-Year Plan stipulated the goal of gradually establishing a carbon-trade market, although the exact details of this were left open. In 2011, the NDRC ordered seven cities and provinces to set up pilot schemes for carbon trading. These were implemented in 2013 and 2014, though the prices of emission rights are very low. Indonesia's own planning calls for energy savings of 17 per cent by 2025. From then on, energy consumption is to grow more slowly than GDP. Brazil introduced programmes on energy conservation and energy efficiency already in 1985 (for electricity) and 1991 (for fossil fuels). Recent government plans aim at a reduction of electricity consumption by 10 per cent by 2030, as compared to a business-as-usual scenario (Ministério de Minas e Energia 2011). Sectoral plans for climate change mitigation have been developed, therein proposing concrete measures to be taken – for instance, by the mining industry and in the transport sector.

Energy-intensive industries in Mexico – such as steel production – are counter-intuitively more energy efficient than in China and the United States; among other reasons, because scrap steel is used, energy-efficient technologies are applied and the share of natural gas as a source of energy for steel plants is relatively high (Rojas Cardenas et al. 2017). This limits the opportunities to save energy by increasing efficiency, and underlines the need for a structural shift in the country's economy. Yet-to-be-built coal-fired power stations will be combined-cycle power plants, which extract energy from waste heat in a second cycle of electricity generation and are thus more efficient than ordinary plants. Methane emissions in oil and gas extraction and natural gas flaring are to be reduced. South Africa has, quite successfully, introduced rapid bus systems in several cities. Eskom's demand-side management programme – a sensible contribution to energy efficiency – has been scaled down, though. A carbon tax is strongly opposed by industry and powerful ministries (Baker et al. 2015). Sector-specific carbon budgets and mitigation strategies remain to be published.

The South African government argues in its INDC that it is "a developing country, with overriding priorities to eliminate poverty and eradicate inequality" (Department of Environmental Affairs 2015: 2). It nevertheless pledges that emissions will peak between 2020 and 2025, stay on this level for up to 10 years and then decline afterwards. Climate Action Tracker rates South Africa's commitment to climate change mitigation as "highly insufficient". The country's emissions target is equivalent to an absolute increase therein by 20 to 82 per cent, as compared to the 1990 level. Under current policies, greenhouse gas emissions will in fact increase by 141 per cent by 2025.

Indonesia has declared intentions to reduce its greenhouse gas emissions by 29 per cent until 2030 (or by 41 per cent, if it receives support from the Global North), as compared to a business-as-usual scenario. This is to be achieved through adaptation in land use and in energy generation, most importantly incorporating a 23 per cent share of renewable energies by 2025. Yet, as noted, coal-fired power generation will expand tremendously. The South East Asian country faces major challenges to limiting emissions from deforestation, or at least their continued rise. Climate Action Tracker rates its commitment as "highly insufficient", especially because of emissions from energy generation likely increasing rapidly up until the early 2030s.

South Korea promises in its INDC to reduce carbon dioxide emissions by 37 per cent by 2030, as compared to business-as-usual scenario (meaning 18 per cent less than in 2010). Operating with rather strong assumptions about GDP growth and a constant share of manufacturing, absolute business-as-usual emissions for 2030 were set rather high – creating the impression that South Korea is willing to make a significant contribution to climate change mitigation. Even if the targets specified in the INDC were met, this would still make for a "highly insufficient" contribution according to Climate Action Tracker. Most problematically, a large share of the reduction commitments – 11 of the aforementioned 37 per cent – is to be achieved through international market mechanisms, whose eventual coming about is highly uncertain. Absolute greenhouse gas emissions by 2030 will most likely be twice as much as in 1990.

India's position on climate change mitigation has changed considerably in the course of this decade. In its INDC, the South Asian state promises to reduce GDP emission intensity by 33 to 35 per cent by 2030 (as compared to the 2005 level), to increase the share of non-fossil electricity generation to 40 per cent of installed capacity by that year and to create a carbon sink of 2.5 to 3 billion tonnes of carbon dioxide equivalent by forest cover. A peak in greenhouse gas emissions is not indicated, meaning that these can increase indefinitely. What is more, meeting the just-mentioned objectives does not require any additional efforts being undertaken. If only the planned capacity addition in renewable energies is achieved, India's emission intensity by 2030 will be less than what the government promised in Paris. Climate Action Tracker, nevertheless, rates India's commitment as "compatible" with limiting global warning to less than 2°C, because the emerging economy is expected to overachieve with regard to the expansion of renewable energies and reduction of emissions intensity.

At the UN Climate Conference in Paris, as well as in previous bilateral negotiations with the United States, the Chinese government agreed that its carbon dioxide emissions will peak by 2030. The carbon intensity of GDP will be reduced by 60 to 65 per cent by that year, as compared to 2005. The share of non-fossil energy sources will rise to 20 per cent. This frame allows for a considerable increase in energy consumption and carbon dioxide emissions in the coming years. Up until 2030, China's energy consumption is expected to increase by 81 to 89 per cent as compared to the 2010 level. Related emissions will rise by 68 to 74 per cent, meaning that the People's Republic alone will most likely absorb nearly 40 per cent of the remaining global carbon budget (Green and Stern 2015; Liu et al. 2015). Climate Action Tracker therefore rates China's commitment to climate change mitigation as "highly insufficient", arguing that while the People's Republic has made some achievements a fair contribution would mean doing significantly more.

Concerning Mexico, it is likely that the liberalisation of the country's energy sector will lead to the large-scale exploitation of unconventional oil and gas resources. By 2040, the emerging economy may produce 3.4 million barrels of oil a day and an annual 15 billion cbm of shale gas (IEA 2016c). Whereas it is somewhat plausible to argue that natural gas replacing coal as a source of electricity constitutes a contribution to climate change mitigation, more oil extraction works against this. In its INDC, Mexico pledges to reduce greenhouse gas emissions by 25 per cent by 2030 (or by 40 per cent, if it receives support from the Global North), as compared to a business-as-usual scenario. According to Climate Action Tracker, Mexico's commitment to climate change mitigation is "insufficient". The predicted emission peak in 2026 is out of sync with the country's present energy policy. The INDC, furthermore, remains to be translated into concrete sector-specific action plans.

According to Brazil's INDC, emissions are to reach 1.2 billion tonnes of carbon dioxide equivalent by 2030 – which equals a 43 per cent reduction as compared to a business-as-usual scenario. Climate Action Tracker argues that the Brazilian commitment is "insufficient", with the Temer conservative government favouring fossil fuels over renewables while deforestation and related emissions have picked up again. With the recent coming to power of Bolsonaro, Brazil's commitment to any form of climate change mitigation has, as noted, now become highly uncertain.

Table C.1 Climate Change Mitigation by the Emerging Economies and Assessment Thereof

	Commitment	*Assessment*	*Main Concerns*
Brazil	emissions reduction by 43 per cent (compared to business as usual)	insufficient	expansion of fossil fuels, increasing deforestation
China	reduction of carbon intensity of GDP by 60 to 65 per cent (compared to 2005), 20 per cent share of non-fossil energy	highly insufficient	vast energy consumption and related emissions, unambitious objectives
India	reduction of GDP emission intensity by 33 to 35 per (compared to 2005), 40 per cent share of non-fossil electricity, additional carbon sinks	compatible	overachievement of renewables expansion, reduction of emissions intensity expected
Indonesia	emissions reduction by 29 per cent, or 41 per cent if supported by Global North (compared to business as usual), 23 per cent share of renewable energies	highly insufficient	rapid expansion of coal sector, few renewables, contribution by land-use uncertain
Mexico	emissions reduction by 25 per cent or 40 per cent if supported by Global North (compared to business as usual)	insufficient	projected emissions peak out of line with current policies (Energy Reform)
South Africa	emissions peak between 2020 and 2025, reduction beginning in 2030s	highly insufficient	enormous increase of emissions, renewable energies delayed, no effective carbon tax
South Korea	emissions reduction by 37 per cent (compared to business as usual)	highly insufficient	strong increase of emissions, over-reliance on international emissions trading schemes

Source: Based on the respective INDCs and assessments by Climate Action Tracker.

FINAL THOUGHTS

In the closing section of this book, we further elaborate on how material conditions, energy demand and expansion plans, politico-economic constellations, energy-related know-how and climate policy all interact with one another. We compare the relevance of different causal factors, and relate our thinking to that of energy transitions and of path dependence. This reveals both the value added by comparative research and also how our approach must be refined going forward.

Energy policy is, first of all, a reflection of material conditions with regard to either the abundance or scarcity of domestic resources. The energy matrices of the emerging economies reflect their respective resource endowments (especially with regard to coal, hydropower and oil and gas). But, as several chapters in this book have shown, energy policy today is rather characterised by a variety of naturally given opportunities than of naturally given constraints. There are, of course, obstacles, such as poor quality of resources and various physical barriers to their exploration. These are not unsurmountable, though. The costs of renewable energies have dropped significantly in the course of the last few years and, depending on how electricity prices are calculated, biomass, solar and wind power are cheaper, cost-competitive or almost cost-competitive compared to fossil fuels. The potential of renewables has been proven by private procurements in Brazil and South Africa, and the rapid state-led expansion of solar and wind energy in China and India. Indonesia, Mexico and South Korea lag behind, but the first two do at least possess considerable naturally given opportunities for the large-scale deployment of renewable energies.

What is striking in this regard is that the emerging economies do not exploit their potential for renewable energy generation to a much greater extent. In other words, material conditions are – as far as they are limited to naturally given conditions – rather helpful to explain the historical trajectories of energy policy that date back to the pre-renewables era but not so much present-day options. The picture changes completely if one takes into consideration man-made material conditions, as these reveal major obstacles to a quick shift to an energy policy that would contribute to climate change mitigation in a meaningful way. First of all, the emerging economies – with the exception of China – each suffer from an inefficient and outdated transmission infrastructure. Pipelines for oil and gas and LNG terminals are few to non-existent. Power grids do not reach all parts of the national territories, and account for high transmission losses. As a consequence, much of the existing potential – in particular for renewables – cannot be exploited. Even China suffers from high curtailment rates for solar and wind power. The energy shortage that marks many of the emerging economies along with increasing prices appear to be, at least partly, a result of infrastructure insufficiencies, rather than of electricity and fossil fuels being in short supply. This also indicates that the focus on creating additional energy-generation capacities – as we have shown, the emerging economies all plan to increase theirs tremendously going forward – is misguided.

Adding another man-made material constraint to an energy transition, numerous power stations that burn coal are currently under construction – and these will remain operational for several decades. Others that have been built since the late 1980s will not be switched off overnight, unless the emerging economies are willing to bear considerable monetary losses on these public investments. Brazil and Mexico have taken steps to expand the extraction of oil and gas. This persistence of fossil fuels is not, as said, due to a lack of alternatives. Using the terminology from energy transition studies, fossil fuels are the dominant socio-technical regime. Their dominant role is reinforced by centralised electricity generation, which is insufficiently flexible to balance against the fluctuations that mark the output of hydropower stations and renewables.

Vested political interests have a similar effect. They range from corruption and self-enrichment (as in South Africa) to the short-term objectives of local and provincial authorities, who stand to lose significantly from decarbonisation while gaining little from an expansion of renewables (as in China, India and Indonesia). As exemplified by China, South Africa and South Korea, the proponents of renewable energies have limited access to government, whereas firms involved in hydrocarbons and nuclear power advance their interests through personal ties to high-ranking politicians – or by virtue of the simple fact that many of these companies are state-owned.

In spite of the persistence of hydrocarbons, we have identified both the critical junctures and forces now putting the currently dominant socio-technical regimes under pressure – thus creating opportunities for a fundamental change of energy policy. The drying-up of fossil resources (both domestically and globally), the escalation of costs to exploit the remaining resources and, therefore, the increasing reliance on energy imports are probably best demonstrated by India, Indonesia and Mexico. They affect, however, all emerging economies except for Brazil, which nevertheless still suffers from frequent electricity price hikes (as do the other emerging economies, too).

It appears, though, that this type of pressure does not lead to either-or decisions being taken. Several coal-fired power stations will be taken off the grid in Mexico and South Africa in the coming years, but these two countries are still building new fossil fuel-based plants while at the same time expanding renewable energy generation. One reason for this is that critical junctures are hardly ever points in time when an all-of-a-sudden and complete change can be made. Our analyses imply that critical junctures open up new pathways that lead to the coexistence of both old and new socio-technical regimes, without the latter replacing the former at once. In this sense, what the emerging economies do is diversify their energy matrices – which, considering the absolute growth of their energy consumption, translates into even more carbon dioxide emissions.

This book has revealed that the emerging economies as polities are characterised by serious efforts to change their energy matrices and energy intensity on the one hand and major obstacles to actually being successful in these regards on the other. All emerging economies pursue impressive objectives on the expansion of renewable energy. They have applied various measures to support renewables, to decrease energy intensity, to tax energy-related emissions and – at least in the cases of China and South Korea – also to switch to a fundamentally different economic model that reflects green growth paradigms. Yet, corruption and institutional deficiencies work against an energy transition. Energy subsidies contradict measures to increase energy efficiency, and their costs have to be carried by power providers themselves – who consequently lack the financial means to invest in energy-related infrastructure (or, indeed, to maintain the existing one). Privatisation has led to mostly positive results, but veto players – from poor citizens to energy-intensive industries – have often opposed it – as well as any significant reduction in energy subsidies. State-owned oil companies in China have successfully pushed back against the transition objectives pursued by the central government. Local and provincial public authorities have obstructed corresponding efforts, not only in the People's Republic but in most of the other emerging economies, too.

As a side remark with regard to political factors, one might expect that democratic systems bring about less environmentally damaging energy policies because democratically elected governments are more dependent on the well-being of their citizens than their counterparts in non-competitive electoral systems are. Moreover, an unrestricted civil society and a free press can reasonably be expected to push for policies that address the environmental concerns of the people. Such expectations are not confirmed by our analysis, however.

On the contrary, China – the only non-democratic emerging economy – has been the most successful country studied in this book in terms of increasing energy efficiency, expanding renewables and advancing an economic paradigm – the new normal – that departs from energy-intensive growth. It, hence, appears that stable autocracies may be more able to implement progressive energy and climate policies because they are less constrained by opposition groups and potential veto players, and therefore highly capable of implementing top-down decisions (Gilley 2012; Ward 2012). This phenomenon is captured by the term "authoritarian environmentalism", which addresses the fact that non-democratic regimes can simply forbid unsustainable behaviour if they wish to (Beeson 2012).

Significant results in terms of energy transitions depend not only on the interplay of relevant political forces but also on technological expertise being available in the countries in question – through acquisition from abroad, domestic innovation and foreign investment. Our analyses have revealed opportunities rather than constraints in this regard. Brazil does currently benefit from its know-how on biofuels, and the rapid expansion of renewables in China has, of course, been facilitated by that country now being the world leader in related fabrication. At the same time, these developments may well have been possible even without biofuels know-how being genuinely Brazilian and capacities for renewables' fabrication being controlled by Chinese firms. Renewable energies have expanded in South Africa, even though the country does not possess any particular technological strengths in this domain. India's wind energy sector is highly attractive to foreign investors – not because of domestic innovation, but because of the potentially huge market presented by this emerging economy. As noted in the introductory chapter, the continuing marginality of renewables in the emerging economies cannot be explained by the former being niche technologies. Their further expansion in these countries is hampered instead by the aforementioned man-made both material and political constraints.

Concluding with a justified and somewhat optimistic observation, there is a striking contrast between the scarcely satisfying INDCs of the emerging economies on the one side and their fairly ambitious objectives vis-à-vis decreasing energy intensity and expanding renewables on the other. To a certain extent, this contrast results from unrealistic objectives being stated in the policy documents that refer to the latter issues. We admit that, as compared to their own expansion plans, the emerging economies tend to overachieve on fossil fuels and underperform with regard to renewables. However, if the lofty ambitions on energy intensity and renewables are at least partly met, some of the emerging economies will eventually make a contribution to climate change mitigation that goes beyond their INDC. One may argue, hence, that the INDCs advance unambitious targets in order to avoid any self-imposed and internationally endorsed restrictions on economic development; also in rejection of any interference by the Global North in the national sovereignty of the emerging economies. Merely national policy documents, meanwhile, are more ambitious, and hopefully closer to what the emerging economies will actually achieve in the coming decades.

NOTE

1. According data is available online at: www.tps-data-portal.org/all-datasets.

References

Acemoglu, Daron et al. 2012. "The Environment and Directed Technical Change". *American Economic Review* 102 (1): 131–66.

Acemoglu, Daron et al. 2016. "Transition to Clean Technology". *Journal of Political Economy* 124 (1): 52–104.

Afonso Nobre, Carlos et al. 2011. "Vulnerability of Brazilian Megacities to Climate Change: The São Paulo Metropolitan Region (RMSP)". In *Climate Change in Brazil: Economic, Social and Regulatory Aspects*, edited by Ronaldo Seroa da Motta et al., 197–220. Brasília: Institute for Applied Economic Research.

Aghion, Philippe et al. 2014. "Path Dependence, Innovation and the Economics of Climate Change". https://www.cccep.ac.uk/wp-content/uploads/2015/10/Aghion_et_al_policy_paper_Nov20141.pdf (13 December 2018).

Agustina, Cut D. et al. 2008. "Black Hole or Black Gold?: The Impact of Oil and Gas Prices on Indonesia's Public Finances". *World Bank Policy Research Working Paper* 4718.

Ahmed, Tofael et al. 2017. "ASEAN Power Grid: A Secure Transmission Infrastructure for Clean and Sustainable Energy for South-East Asia". *Renewable and Sustainable Energy Reviews* 67: 1420–35.

Ahn, Sun J., and Dagmar Graczyk. 2012. "Understanding Energy Challenges in India: Policies, Players and Issues". https://www.iea.org/publications/freepublications/publication/India_study_FINAL_WEB.pdf (1 May 2018).

Aldy, Joseph E., William A. Pizer and Keigo Akimoto. 2017. "Comparing Emissions Mitigation Efforts across Countries". *Climate Policy* 17 (4): 501–15.

Alpizar, Israel, and Carlos Rodríguez. 2016. "Review of Mexico's Energy Reform in 2013: Background, Analysis of the Reform and Reactions". *Renewable and Sustainable Energy Review* 58: 725–36.

Alstom. 2011. "Alstom to Invest in a Global Technology Centre for Hydro Power Products in Brazil". http://www.alstom.com/press-centre/2011/11/alstom-to-invest-global-technology-centre-hydro-power-products-brazil (8 January 2018).

Alstom. 2014. "Alstom Inaugurates Its First Hydropower Global Technology Center in Latin America". http://www.alstom.com/press-centre/2014/5/alstom-inaugurates-its-first-hydropower-global-technology-center-in-latin-america (8 January 2018).

Alvarez, Jorge, and Fabian Valencia. 2016. "Made in Mexico: Energy Reform and Manufacturing Growth". *Energy Economics* 55 (3): 253–65.

Alves, Ana C. 2015. "China and Brazil in Sub-Saharan African Fossil Fuels: A Comparative Analysis". In *A New Scramble for Africa?: The Rush for Energy Resources in Sub-Saharan Africa*, edited by Sören Scholvin, 33–51. Farnham: Ashgate.

Alyousef, Yousef, and Abdulhadi Varnham. 2010. "Saudi Arabia's National Energy Efficiency Programme: Description, Achievements and Way Forward". *International Journal of Low-Carbon Technologies* 5 (4): 291–97.

Anand, Madhav R., and Dev N. Prasad. 2013. "Recent Trends in Production and Import of Coal in India". *Ministry of Coal Occasional Working Paper* 1/2013.

Andersen, Lykke E. et al. 2016. *Climate Change Impacts and Household Resilience: Prospects for 2050 in Brazil, Mexico, and Peru.* Washington, DC: IFPRI.

Andrews-Speed, Philip. 2016. "Connecting ASEAN through the Power Grid: Next Steps". *Energy Studies Institute Policy Brief* 11/2016.

Andrews-Speed, Philip et al. 2014. "China's Energy Crossroads: Forging a New Energy and Environmental Balance". http://nbr.org/publications/specialreport/pdf/sr47_chinas_energy_crossroads_november2014.pdf (26 April 2018).

Andrews-Speed, Philip, and Xunpeng Shi. 2016. "What Role Can the G20 Play in Global Energy Governance?: Implications for China's Presidency". *Global Policy* 7 (2): 198–206.

Asian Development Bank. 2016. "Indonesia – Energy Sector: Assessment, Strategy and Roadmap". https://www.adb.org/sites/default/files/institutional-document/189713/ino-energy-asr.pdf (10 March 2017).

Bagley, Bruce M., Dina Moulioukova and Hanna S. Kassab (eds). 2015. *The Impact of Emerging Economies on Global Energy and the Environment: Challenges Ahead.* Lanham, MD: Lexington.

Baker, Lucy. 2015. "Renewable Energy in South Africa's Minerals–Energy Complex: A 'Low Carbon' Transition?". *Review of African Political Economy* 144: 245–61.

Baker, Lucy, Peter Newell and Jon Phillips. 2014. "The Political Economy of Energy Transitions: The Case of South Africa". *New Political Economy* 19 (6): 791–818.

Baker, Lucy et al. 2015. "The Political Economy of Decarbonisation: Exploring the Dynamics of South Africa's Electricity Sector". http://www.erc.uct.ac.za/sites/default/files/image_tool/images/119/Papers-2015/15-Baker-etal-Political_economy_decarbonisation.pdf (4 May 2018).

Barta, Bo. 2011. "Renewable Energy: Hydropower". In *Sustainable Energy Resource Handbook: South Africa 2*, edited by Erik Kiderlen, Barry Bredenkamp and Lauren Smith, 139–51. Cape Town: Alive2Green.

Bath, Elizabeth et al. 2015. "Empty Promise: G20 Subsidies to Oil, Gas and Coal Production". https://www.odi.org/sites/odi.org.uk/files/odi-assets/publications-opinion-files/9957.pdf (29 April 2018).

Baumgartner, Daniel, and Nadine Godehardt. 2012. "Chinas Energiepolitik und die Strategien der Nationalen Ölkonzerne". *GIGA Focus* 3/2012.

Beaton, Christopher, Lucky Lontoh and Matthew Wai-Poi. 2017. "Indonesia: Pricing Reforms, Social Assistance, and the Importance of Perceptions". In *The Political Economy of Energy Subsidy Reform*, edited by Gabriela Inchauste and David G. Victor, 133–207. Washington, DC: World Bank.

Beeson, Mark. 2010. "The Coming of Environmental Authoritarianism". *Environmental Politics* 19 (2): 276–94.

Béguin, Andreas et al. 2011. "The Opposing Effects of Climate Change and Socio-Economic Development on the Global Distribution of Malaria". *Global Environmental Change* 21 (4): 1209–14.

Benhin James K. 2006. "Climate Change and South African Agriculture". *CEEPA Discussion Paper* 21.

Bentley, Roger W. 2002. "Global Oil and Gas Depletion: An Overview". *Energy Policy* 30 (3): 189–205.

Betz, Joachim. 2012. "India's Turn in Climate Policy: Assessing the Interplay of Domestic and International Policy Change". *GIGA Working Paper* 190.

Bischof-Niemz, Tobias, and Terence Creamer. 2018. *South Africa's Energy Transition: A Roadmap to a Decarbonised, Low-Cost and Job-Rich Future*. London: Routledge.

Blignaut, James. 2012. "Climate Change: The Opportunity Cost of Medupi and Kusile Power Stations". *Journal of Energy in Southern Africa* 23 (4): 67–75.

Bonafé, Ernesto, Ernesto Beltrán and Juan F. Neira. 2016. "Mexico's Energy Sector under the Universal Principles of the 2015 International Energy Charter". http://www.energycharter.org/fileadmin/DocumentsMedia/Occasional/Mexico_Report.pdf (6 September 2017).

BP. 2014. "Energy Outlook 2035". http://www.bp.com/content/dam/bp/pdf/Energy-economics/energy-outlook-2015/Energy_Outlook_2035_booklet.pdf (24 April 2015).

BP. 2016. "BP Statistical Review of World Energy". https://www.bp.com/content/dam/bp/en/corporate/pdf/energy-economics/statistical-review-2017/bp-statistical-review-of-world-energy-2017-full-report.pdf (29 April 2018).

Braithwaite, David et al. 2012. "Indonesia's Fuel Subsidies: Action Plan for Reform". http://www.iisd.org/gsi/sites/default/files/ffs_actionplan_indonesia.pdf (16 February 2017).

Bravo, Natalie, and Nikos Tsafos. 2015. "Indonesia: The Nexus of Gas and Electricity". In *Indonesia: A Regional Energy Leader in Transition*, edited by Natalie Bravo et al., 33–45. Washington, DC: NBR.

Brown, Jessica, and Leo Peskett. 2011. "Climate Finance in Indonesia: Lessons for the Future of Public Finance". *ODI Working Paper* 11.

Buckley, Tim. 2015. "India's Electricity-Sector Transformation". http://ieefa.org/wp-content/uploads/2015/08/IEEFA-Indian-Electricity-Sector-Transformation-11-August-2015.pdf (7 June 2018).

Buckley, Tim, and Tom Sanzillo. 2015. "Past Peak Coal in China". http://ieefa.org/wp-content/uploads/2015/11/IEEFA_Peak-Coal_November-2015.pdf (26 April 2018).

Buckley, Tim, and Jai Sharda. 2016. "Data Bite: December Numbers Signal a Looming End to India's Appetite for Foreign Coal". http://ieefa.org/ieefa-commentary-december-data-from-india-signals-a-looming-end-to-its-coal-import-market (1 May 2018).

Bueno Pereira, Enio et al. 2006. "Atlas Brasileiro de Energia Solar". http://ftp.cptec.inpe.br/labren/publ/livros/brazil_solar_atlas_R1.pdf (12 April 2017).

Cámara de Diputados [of Mexico]. 2016. "Ley General de Cambio Climático". http://www.diputados.gob.mx/LeyesBiblio/pdf/LGCC_010616.pdf (30 August 2017).

Camargo do Amarante, Odilon A. et al. 2001. "Atlas do Potencial Eólico Brasileiro". http://www.cresesb.cepel.br/publicacoes/download/atlas_eolico/Atlas%20do%20Potencial%20Eolico%20Brasileiro.pdf (4 January 2017).

Camilleri, James. 2011. "India's Insatiable Appetite". https://www.academia.edu/801985/Indias_Insatiable_Appetite (23 May 2018).

Caprioni, Elena. 2013. "Why Are Foreign Mining Companies Retreating from China?". http://www.asiapacificmemo.ca/why-are-foreign-mining-companies-retreating-from-china (24 April 2018).

Carbon Tracker Initiative. 2014. "The Great Coal Cap: China's Energy Policies and the Financial Implications for Thermal Coal". https://www.carbontracker.org/reports/the-great-coal-cap-chinas-energy-policies-and-the-financial-implications-for-thermal-coal (26 April 2018).

Carvalho, Joaquim F., and Ildo L. Sauer. 2009. "Does Brazil Need New Nuclear Power Plants?". *Energy Policy* 37 (1): 580–84.

Central Electricity Authority [of India]. 2016a. "All India Installed Capacity (in MW) of Power Stations". http://www.cea.nic.in/reports/monthly/installedcapacity/2017/installed_capacity-01.pdf (1 May 2018).

Central Electricity Authority [of India]. 2016b. "Executive Summary: Power Sector". http://cea.nic.in/reports/monthly/executivesummary/2016/exe_summary-04.pdf (1 May 2018).

Charles, Roger, David Held and Eva-Maria Nag. 2013. "Controlling the Amazon: Brazil's Evolving Response to Climate Change". In *Climate Governance in the Developing World*, edited by David Held, 134–54, Cambridge: Polity Press.

Chatterjee, Anidita, and Mohit Sinha. 2006. *Power Realignments in Asia: China, India and the United States, Report of a US-India Policy Dialogue*. Philadelphia: CASI.

Cheung, Kat. 2011. "Integration of Renewables: Status and Challenges in China". https://www.iea.org/publications/freepublications/publication/Integration_of_Renewables.pdf (26 April 2018).

Choi, Sang D. 2015. "The Green Growth Movement in the Republic of Korea: Option or Necessity?". http://documents.worldbank.org/curated/en/145991468000018642/pdf/101569-BRI-PUBLIC-ADD-SERIES-Box394816B.pdf (29 April 2018).

Chung, Woo J. 2014. "Update on ROK Energy Sector and Energy Policies". http://nautilus.org/wp-content/uploads/2013/08/ROK_Chung_Update-ROK-energy-sector-and-policies.pdf (29 April 2018).

Climate Transparency. 2016. "Brown to Green. Assessing the G20 Transition to a Low-Carbon Economy". http://www.climate-transparency.org/wp-content/

uploads/2016/08/Brown-to-Green-Assessing-the-G20-transition-to-a-low-carbon-economy.pdf (4 June 2018).
Coady, David et al. 2015. "How Large Are Global Energy Subsidies?". *IMF Working Paper* 15/105.
Cohen, Brett, and Harald Winkler. 2014. "Greenhouse Gas Emissions from Shale Gas and Coal for Electricity Generation in South Africa". *South African Journal of Science* 110 (3–4): n.p.
Collier, Ruth B., and David Collier. 1991. *Shaping the Political Arena: Critical Junctures, the Labor Movement, and Regime Dynamics in Latin America*. Princeton, NJ: Princeton University Press.
Compston, Hugh, and Ian Bailey. 2016. "Climate Policy Strength Compared: China, the US, the EU, India, Russia, and Japan". *Climate Policy* 16 (2): 145–64.
Cornot-Gandolphe, Sylvie. 2018. "South Korea's New Electricity Plan: Cosmetic Changes or a Breakthrough for the Climate?". https://www.ifri.org/sites/default/files/atoms/files/cornotgandolphe_south_korea_electricity_2018.pdf (12 December 2018).
COSIPLAN. 2016. "Cartera de Proyectos 2016". https://www.flipsnack.com/IIRSA/informe-de-la-cartera-de-proyectos-del-cosiplan-2016.html (24 March 2017).
Council on Climate Change [of India]. 2007. "National Action Plan on Climate Change". http://www.moef.nic.in/downloads/home/Pg01-52.pdf (1 May 2018).
Cuesta Fernández, Ivan. 2015. "Mammoth Dams, Lean Neighbours: Assessing the Bid to Turn Ethiopia into East Africa's Powerhouse". In *A New Scramble for Africa?: The Rush for Energy Resources in Sub-Saharan Africa*, edited by Sören Scholvin, 93–110. Farnham: Ashgate.
Cunningham, Nick. 2016. "Record Loss for Petrobras as Political and Economic Crisis Worsen". https://oilprice.com/Energy/Crude-Oil/Record-Loss-For-Petrobras-As-Political-And-Economic-Crisis-Worsen.html (8 January 2018).
Das, Anjana, and Erik O. Ahlgren. 2010. "Implications of Using Clean Technologies to Power Selected ASEAN Countries". *Energy Policy* 38 (4): 1851–71.
Dave, Bhavna. 2016. "Resetting India's Engagement in Central Asia: From Symbols to Substance". https://www.rsis.edu.sg/wp-content/uploads/2016/02/PR160202_Resetting-Indias-Engagment.pdf (13 December 2018).
David, Paul. 1986. "Understanding the Economics of QWERTY: The Necessity of History". In *Economic History and the Modern Economist*, edited by William N. Parker, 32–46. New York: Blackwell.
Deheza, Elizabeth, and Jorge Mora. 2013. "Climate Change, Migration and Security: Best-Practice Policy and Operational Options for Mexico". *Whitehall Report* 1/2013.
Department of Energy [of South Africa]. 2011. "Integrated Resource Plan for Electricity 2010–2030". http://www.doe-irp.co.za/content/IRP2010_promulgated.pdf (11 January 2017).
Department of Energy [of South Africa]. 2013. "Integrated Resource Plan for Electricity 2010–2030: Update Report". http://www.doe-irp.co.za/content/IRP2010_updatea.pdf (11 January 2017).

Department of Energy [of South Africa]. 2016. "Integrated Resource Plan Update: Assumptions, Base Case Results and Observations". http://www.energy.gov.za/IRP/2016/Draft-IRP-2016-Assumptions-Base-Case-and-Observations-Revision1.pdf (11 January 2017).

Department of Environmental Affairs [of South Africa]. 2011. "National Climate Change Response White Paper". https://www.environment.gov.za/sites/default/files/legislations/national_climatechange_response_whitepaper.pdf (21 January 2017).

Department of Environmental Affairs [of South Africa]. 2014. "GHG Inventory for South Africa: 2000–2010". https://www.environment.gov.za/sites/default/files/docs/greenhousegas_invetorysouthafrica.pdf (20 January 2017).

Department of Environmental Affairs [of South Africa]. 2015. "Intended Nationally Determined Contribution". http://www4.unfccc.int/ndcregistry/PublishedDocuments/South%20Africa%20First/South%20Africa.pdf (21 January 2017).

Department of Finance [of South Africa]. 1996. "Growth, Employment and Redistribution: A Macroeconomic Strategy". http://www.treasury.gov.za/publications/other/gear/chapters.pdf (9 January 2018).

Diamond, Jared. 2005. *Collapse: How Societies Choose to Fail or Succeed*. New York: Viking.

Dorian, James P., Bratt Wigortz and Dru Gladney. 1997. "Central Asia and Xinjiang, China: Emerging Energy, Economic and Ethnic Relations". *Central Asian Survey* 16 (4): 461–86.

Dorian, James P. et al. 1999. "Energy in Central Asia and Northwest China: Major Trends and Opportunities for Regional Cooperation". *Energy Policy* 27 (5): 281–97.

Downs, Erica S. 2004. "The Chinese Energy Security Debate". *China Quarterly* 177: 21–41.

Downs, Erica S. 2008. "Business Interest Groups in Chinese Politics: The Case of Oil Companies". In *China's Changing Political Landscape: Prospects for Democracy*, edited by Cheng Li, 121–41. Washington, DC: Brookings.

Dubash, Navroz K. (ed.). 2015. *Handbook of Climate Change and India: Development, Politics and Governance*. London: Routledge.

Duffield, John S. 2014. "South Korea's National Energy Plan Six Years On". *Asian Politics & Policy* 6 (3): 433–54.

Dupont, Claire, and Sebstatian Oberthür (eds). 2015. *Decarbonization in the European Union: Internal Policies and External Strategies*. London: Palgrave Macmillan.

Earthlife. 2009. "Protest against Sasol's CMD Project". http://earthlife.org.za/2009/02/25/protest-against-sasols-cdm-project (10 January 2018).

Eberhard, Anton, Joel Kolker and James Leigland. 2014. "South Africa's Renewable Energy IPP Procurement Program: Success Factors and Lessons". http://www.gsb.uct.ac.za/files/ppiafreport.pdf (21 January 2017).

Eberhard, Anton, and Raine Naude. 2016. "The South African Renewable Energy Independent Power Producer Procurement Programme: A Review and Lessons Learned". *Journal of Energy in Southern Africa* 27 (4): 1–14.

Ebinger, Charles K. 2016. "India's Energy and Climate Policy: Can India Meet the Challenge of Industrialization and Climate Change?". https://www.brookings.edu/wp-content/uploads/2016/07/india_energy_climate_policy_ebinger.pdf (1 May 2018).

Ecofys. 2016. "International Comparison of Fossil Fuel Power Efficiency and CO_2 Intensity: Update 2016". https://www.ecofys.com/files/files/intern.-comparison-of-fossil-power-efficiency-and-co2-intensity-2016.pdf (29 April 2018).

Economic Development Department [of South Africa]. 2010. "The New Growth Path: The Framework". http://www.info.gov.za/view/DownloadFileAction?id=135748 (11 December 2017).

Edkins, Max, Andrew Marquard and Harald Winkler. 2010. "Assessing the Effectiveness of National Solar and Wind Energy Policies in South Africa". https://open.uct.ac.za/bitstream/item/19472/Edkins_Assessing_effectiveness_2010.pdf (4 May 2018).

Edney, Kingsley, and Jonathan Symons. 2014. "China and the Blunt Temptations of Geo-Engineering: The Role of Solar Radiation Management in China's Strategic Response to Climate Change". *Pacific Review* 27 (3): 307–32.

EIA. 2013a. "Technically Recoverable Shale Oil and Shale Gas Resources: Brazil". https://www.eia.gov/analysis/studies/worldshalegas/pdf/Brazil_2013.pdf (4 January 2017).

EIA. 2013b. "Technically Recoverable Shale Oil and Shale Gas Resources: An Assessment of 137 Shale Formations in 41 Countries outside the United States". http://www.eia.gov/analysis/studies/worldshalegas/pdf/fullreport.pdf (22 November 2016).

EIA. 2014a. "International Energy Outlook 2014". http://www.eia.gov/forecasts/ieo/pdf/0484(2014).pdf (24 April 2015).

EIA. 2015a. "International Energy Statistics: Total Primary Energy Consumption". http://www.eia.gov/cfapps/ipdbproject/iedindex3.cfm?tid=44&pid=45&aid=2&cid=regions&syid=2005&eyid=2009&unit=QBTU (26 April 2015).

EIA. 2015b. "International Energy Statistics: Total Primary Energy Consumption per Dollar of GDP". http://www.eia.gov/cfapps/ipdbproject/iedindex3.cfm?tid=92&pid=47&aid=2&cid=regions&syid=2011&eyid=2011&unit=BTUPUSDM (26 April 2015).

EIA. 2015c. "International Energy Statistics: Total Carbon Dioxide Emissions from the Consumption of Energy". http://www.eia.gov/cfapps/ipdbproject/iedindex3.cfm?tid=90&pid=44&aid=8&cid=regions&syid=2012&eyid=2012&unit=MMTCD (26 April 2015).

EIA. 2015d. "China". https://www.eia.gov/beta/international/analysis_includes/countries_long/China/china.pdf (24 January 2018).

EIA. 2015e. "Indonesia". https://www.eia.gov/beta/international/analysis_includes/countries_long/Indonesia/indonesia.pdf (22 June 2017).

EIA. 2015f. "Brazil". https://www.eia.gov/beta/international/analysis_includes/countries_long/Brazil/brazil.pdf (21 April 2016).

EIA. 2015g. "South Africa". https://www.eia.gov/beta/international/analysis_includes/countries_long/South_Africa/south_africa.pdf (3 January 2017).

EIA. 2016a. "India". https://www.eia.gov/beta/international/analysis_includes/countries_long/India/india.pdf (24 January 2018).
EIA. 2016b. "Mexico". https://www.eia.gov/beta/international/analysis_includes/countries_long/Mexico/mexico.pdf (19 July 2017).
EIA. 2017. "South Korea". https://www.eia.gov/beta/international/analysis_includes/countries_long/Korea_South/south_korea.pdf (29 April 2018).
Enerdata. 2017. "Global Energy Statistical Yearbook 2017". https://yearbook.enerdata.net (29 April 2018).
Energy Research Centre. 2013. "Towards a New Power Plan". Report prepared for the National Planning Commission of South Africa, in possession of the authors.
Energy Research Centre. 2015. "South Africa's Proposed Nuclear Build Plan: An Analysis of the Potential Socioeconomic Risks". http://www.erc.uct.ac.za/sites/default/files/image_tool/images/119/Papers-2015/15-ERC-Nuclear_build_plan_Technical_report.pdf (4 May 2018).
Energy Watch Group. 2007. "Coal: Resources and Future Production". http://energywatchgroup.org/wp-content/uploads/2014/02/EWG_Report_Coal_10-07-2007ms1.pdf (24 April 2015).
Energy Watch Group. 2013. "Fossil and Nuclear Fuels: The Supply Outlook". http://energywatchgroup.org/wp-content/uploads/2014/02/EWG-update2013_long_18_03_2013up1.pdf (24 April 2015).
ENGO Network on CCS. 2015. "Closing the Gap on Climate: Why CCS Is a Vital Part of the Solution". http://hub.globalccsinstitute.com/sites/default/files/publications/197903/closing-gap-climate-ccs-vital-part-solution.pdf (26 April 2018).
Ernst & Young. 2011. "India's Energy Security: Key Issues Impacting the Indian Oil and Gas Sector". http://www.ey.com/Publication/vwLUAssets/Indias_energy_security/$FILE/India-s_energy_security.pdf (1 May 2018).
Ernst & Young. 2013. "Mapping India's Renewable Energy Growth Potential: Status and Outlook 2013". http://www.ey.com/Publication/vwLUAssets/Mapping_Indias_Renewable_Energy_growth_potential/%24FILE/EY-Mapping-Indias-Renewable-Energy-growth-potential.pdf (1 May 2018).
Eskom. 2008. "Proposed Waterberg Coal Fired Power Stations: Strategic Overview". http://www.eskom.co.za/OurCompany/SustainableDevelopment/EnvironmentalImpactAssessments/Documents/Ann_D3c_Eskom_Presentation_App_21.pdf (10 January 2018).
Eskom. 2015. "Integrated Report". http://www.eskom.co.za/IR2015/Documents/EskomIR2015single.pdf (22 November 2016).
Eskom. n.d. "Historical Average Price Increase". http://www.eskom.co.za/CustomerCare/TariffsAndCharges/Documents/Historical%20average%20prices%20and%20increase_v20179821.xlsx (9 January 2018).
Fargione, Joseph et al. 2008. "Land Clearing and the Biofuel Carbon Debt". *Science* 5867: 1235–38.
Federative Republic of Brazil. n.d.. "Intended Nationally Determined Contribution". http://www4.unfccc.int/submissions/INDC/Published%20Documents/Brazil/1/BRAZIL%20iNDC%20english%20FINAL.pdf (20 June 2017).
Feng, Shuaizhang, Alan B. Krueger and Michael Oppenheimer. 2010. "Linkages among Climate Change, Crop Yields and Mexico–US Cross-Border Migration".

Proceedings of the National Academy of Sciences of the United States of America 107 (32): 14257–62.

Fig, David, and Sören Scholvin. 2015. "Fracking the Karoo: The Barriers to Shale Gas Extraction in South Africa Based on Experiences from Europe and the US". In *A New Scramble for Africa?: The Rush for Energy Resources in Sub-Saharan Africa*, edited by Sören Scholvin, 131–47. Farnham: Ashgate.

Fine, Ben. 2008. "The Minerals–Energy Complex Is Dead: Long Live the MEC?". Paper presented at the Amandla Colloquium on Continuity and Discontinuity of Capitalism in the Post-Apartheid South Africa, Cape Town, 4–6 April.

Fine, Ben, and Zavareh Rustomjee. 1996. *The Political Economy of South Africa: From Minerals-Energy Complex to Industrialisation*. London: Hurst.

Fluri, Thomas P., and Theodor W. von Backström. 2009. "Comparison of Solar Resource Assessments for South Africa". In *Proceedings of the ISES Solar World Congress 2009: Renewable Energy Shaping Our Future*, edited by International Solar Energy Society, 68–76. Johannesburg: ISES.

Fouquet, Roger. 2010. "The Slow Search for Solutions: Lessons from Historical Energy Transitions by Sector and Service". *Energy Policy* 38 (11): 6586–96.

Foxon, Timothy J. 2011. "A Coevolutionary Framework for Analyzing a Transition to a Sustainable Low Carbon Economy". *Ecological Economics* 70 (12): 2258–67.

Foxon, Timothy J. et al. 2005. "Transforming Policy Processes to Promote Sustainable Innovation: Some Guiding Principles: A Report for Policy Makers". http://www.sustainabletechnologies.ac.uk/PDF/project%20reports/SI_policy_guidance_final_version.pdf (11 December 2017).

Foxon, Timothy J., Jonathan H. Köhler and Christine Oughton. 2008. *Innovation for a Low Carbon Economy: Economic, Institutional and Management Approaches*. Cheltenham: Elgar.

Frank, Charles. 2016. "India: Potential for Even Greater Emission Reduction". https://www.brookings.edu/wp-content/uploads/2016/08/global_20160818_cop21_india.pdf (1 May 2018).

FS-UNEP and BNEF. 2016. "Global Trends in Renewable Energy Investment 2016". http://fs-unep-centre.org/sites/default/files/publications/globaltrendsinrenewableenergyinvestment2016lowres_0.pdf (25 April 2018).

Galindo, Luis M. 2010. *The Economics of Climate Change in Mexico*. Mexico City: Semarnat.

GDP MAPS Working Group. 2016. "Burden of Disease Attributable to Coal-Burning and Other Major Sources of Air Pollution in China". https://www.healtheffects.org/system/files/GBDMAPS-ReportEnglishFinal1.pdf (24 January 2018).

Geels, Frank W. 2002. "Technological Transitions as Evolutionary Reconfiguration Processes: A Multi-Level Perspective and a Case Study". *Research Policy* 318–19: 1257–74.

Geels, Frank W. 2014. "Regime Resistance against Low-Carbon Transitions: Introducing Politics and Power into the Multi-Level Perspective". *Theory, Culture & Society* 31 (5): 21–40.

Geels, Frank W., and Johan Schot. 2007. "Typology of Sociotechnical Transition Pathways". *Research Policy* 36 (3): 399–417.

Gilley, Bruce. 2012. "Authoritarian Environmentalism and China's Response to Climate Change". *Environmental Politics* 21 (2): 287–307.

Global CCS Institute. 2013. "India CCS Scoping Study: Final Report". http://hub.globalccsinstitute.com/sites/default/files/publications/88981/india-ccs-scoping-study-final-report.pdf (1 May 2018).

Global Green Growth Institute. 2015. "Korea's Green Growth Experience: Process, Outcomes and Lessons Learned". http://www.greengrowthknowledge.org/sites/default/files/downloads/resource/Koreas-Green-Growth-Experience_GGGI.pdf (29 April 2018).

Goerild, Shailly K. et al. 2014. *Low Carbon Development in China and India: Issues and Strategies*. New Delhi: TERI.

Gomes, Ieda. 2014. "Brazil: Country of the Future or Has Its Time Come for Natural Gas?". *OIES Paper* 88.

Gosens, Jorrit, and Yonglong Lu. 2014. "Prospects for Global Market Expansion of China's Wind Turbine Manufacturing Industry". *Energy Policy* 67: 301–18.

Government of India. 2015. "Intended Nationally Determined Contribution". http://www4.unfccc.int/ndcregistry/PublishedDocuments/India%20First/INDIA%20INDC%20TO%20UNFCCC.pdf (1 May 2018).

Government of India. 2016. "State of the Economy: An Overview". http://www.indiaenvironmentportal.org.in/files/file/Economic%20Survey%202015%E2%80%9316%20%20-%20vol%202.pdf (1 May 2018).

Governo Federal [of Brazil]. 2007. "Plano Nacional sobre Mudança do Clima". http://www.mma.gov.br/estruturas/smcq_climaticas/_arquivos/plano_nacional_mudanca_clima.pdf (20 June 2017).

Green, Fergus, and Nicholas Stern. 2015. "China's 'New Normal': Structural Change, Better Growth, and Peak Emissions". http://www.lse.ac.uk/GranthamInstitute/wp-content/uploads/2015/06/China_new_normal_web1.pdf (25 April 2018).

Green, Fergus, and Nicholas Stern. 2017. "China's Changing Economy: Implications for Its Carbon Dioxide Emissions". *Climate Policy* 17 (4): 423–42.

Greenberg, Stephen. 2009. "Market Liberalisation and Continental Expansion: The Repositioning of Eskom in post-Apartheid South Africa". In *Electric Capitalism: Recolonising Africa on the Power Grid*, edited by David A. McDonald, 73–108. London: Earthscan.

Group Centennial. 2009. "India 2039: Unleashing an Energy Revolution". *Asian Development Bank Policy Paper* 4/2009.

Gu, Xuewu. 2011. "China: Energiepartner oder Energiekonkurrent im 21. Jahrhundert?". Speech given at the BP Forum, Berlin, 30 November.

Guevara, Zeus et al. 2017. "The Status and Evolution of Energy Supply and Use in Mexico Prior to the 2014 Energy Reform: An Input-Output Approach". *Economies* 5 (1): n.p.

Gutiérrez Negrín, Luis C. 2012. "Update of the Geothermal Electric Potential in Mexico". *Geothermal Resources Council Transactions* 36: 671–77.

Günther, Stephan. 2005. "Ein altbekannter Teufelskreis: Chinas Dominanz im Süd-Süd-Handel erzeugt neue Abhängigkeiten". *iz3w* 287: 25–27.

Gunningham, Neil. 2014. "A Shale Gas Revolution for China?". *Climate Policy* 14 (2): 302–20.

Gupta, Arvind. 2013. "India and Central Asia: Need for a Pro-Active Approach". https://idsa.in/system/files/PB_IndiaandCentralAsia.pdf (13 December 2018).

Gupta, Rajan, Harihar Shankar and Sunjoy Joshi. 2012. "Development, Energy Security and Climate Security: India's Converging Goals". http://globalenergyobservatory.org/docs/analysis_papers/Gupta_ORF_Conf_final(v10).pdf (1 May 2018).

Hallding, Karl, Guoyi Han and Marie Olsson. 2009. "A Balancing Act: China's Role in Climate Change". http://www.regeringen.se/49b6b8/contentassets/ddb77a1cb0bb441dbe2981b2a019253a/a-balancing-act-chinas-role-in-climate-change-rapport (25 April 2018).

Hein, Jonas. 2013. "Climate Change Mitigation in Emerging Economies: The Case of Indonesia – Hot Air or Leadership?". *DIE Briefing Paper* 8/2013.

Held, David (ed.). 2013. *Climate Governance in the Developing World*. Cambridge: Polity Press.

Hernández Escobedo, Quetzalcóatl et al. 2014. "Wind Energy Resource in Northern Mexico". *Renewable and Sustainable Energy Reviews* 32: 890–914.

Hilton, Isabel, and Oliver Kerr. 2017. "The Paris Agreement: China's 'New Normal' Role in International Climate Negotiations". *Climate Policy* 17 (1): 248–58.

Hughes, David J. 2013. *Drill, Baby, Drill: Can Unconventional Fuels Usher in a New Era of Energy Abundance*. Santa Rosa, CA: Post Carbon Institute.

Husar, Jörg, and Dennis Best. 2013. "Energy Investments and Technology Transfer Across Emerging Economies: The Case of Brazil and China". https://www.iea.org/publications/freepublications/publication/PCS_ChinaBrazil_FINAL_WEB.pdf (19 June 2017).

Ibrahim, Herman D. 2014. "Indonesia Energy Projection and Energy Policy Toward 2050". Presentation at the Smart Grid Seminar, Netherlands Energy Mission to Indonesia, Jakarta, 27 January.

IEA. 2007. "World Energy Outlook 2007: China and India Insights". http://www.worldenergyoutlook.org/media/weowebsite/2008-1994/weo_2007.pdf (1 May 2018).

IEA. 2009. "World Energy Outlook 2009". http://www.worldenergyoutlook.org/media/weowebsite/2009/WEO2009.pdf (25 April 2018).

IEA. 2010. "World Energy Outlook 2010". http://www.worldenergyoutlook.org/media/weo2010.pdf (25 April 2018).

IEA. 2011. "Technology Development Prospects for the Indian Power Sector". http://www.iea.org/publications/freepublications/publication/technology_development_india.pdf (21 July 2015).

IEA. 2012. "Energy Policies of IEA Countries: The Republic of Korea, 2012 Review". https://www.iea.org/publications/freepublications/publication/Korea2012_free.pdf (29 April 2018).

IEA. 2013. "World Energy Outlook 2013". http://www.iea.org/publications/freepublications/publication/WEO2013.pdf (26 April 2016).

IEA. 2014a. "World Energy Outlook 2014". https://www.iea.org/publications/freepublications/publication/WEO2014.pdf (25 April 2018).

IEA. 2014b. “Update on Overseas Investments by China’s National Oil Companies: Achievements and Challenges since 2011”. https://www.iea.org/publications/freepublications/publication/PartnerCountrySeriesUpdateonOverseasInvestmentsbyChinasNationalOilCompanies.pdf (25 April 2018).

IEA. 2014c. “Energy Supply Security: Emergency Response of IEA Countries 2014”. https://www.iea.org/publications/freepublications/publication/ENERGYSUPPLYSECURITY2014.pdf (29 April 2018).

IEA. 2015a. “World Energy Outlook 2015”. https://www.iea.org/publications/freepublications/publication/WEO2015.pdf (25 April 2018).

IEA. 2015b. “Indonesia 2015”. https://www.iea.org/publications/freepublications/publication/Indonesia_IDR.pdf (11 February 2017).

IEA. 2016a. *World Energy Outlook 2016*. Paris: IEA.

IEA. 2016b. “Energy Efficiency Market Report 2016”. https://www.iea.org/eemr16/files/medium-term-energy-efficiency-2016_WEB.PDF (25 April 2018).

IEA. 2016c. “Mexico Energy Outlook”. https://www.iea.org/publications/freepublications/publication/MexicoEnergyOutlook.pdf (19 July 2017).

IEA. 2016d. “Fossil Fuel Subsidy Reform in Mexico and Indonesia”. https://www.iea.org/publications/freepublications/publication/PartnerCountrySeriesFossil_Fuel_Subsidy_Reform_Mexico_Indonesia_2016_WEB.pdf (19 July 2017).

IEA. 2017. *World Energy Outlook 2017*. Paris: IEA.

Inman, Mason. 2014. “Natural Gas: The Fracking Fallacy”. *Nature* 516: 28–30.

In-Ha, Hwang. 2014. “South Korea’s National Basic Plan for New and Renewable Energies”. https://eneken.ieej.or.jp/data/5842.pdf (29 April 2018).

IPCC. 2014. “Fifth Assessment Report”. https://www.ipcc.ch/report/ar5 (23 May 2018).

IRENA. 2014. “Renewable Energy Prospects: China”. http://www.irena.org/-/media/Files/IRENA/Agency/Publication/2014/IRENA_REmap_China_report_2014.pdf (25 April 2018).

IRENA. 2015a. “Renewable Energy Prospects: Mexico”. http://www.irena.org/DocumentDownloads/Publications/IRENA_REmap_Mexico_report_2015.pdf (19 July 2017).

IRENA. 2015b. “Renewable Energy Policy Brief: Mexico”. http://www.irena.org/DocumentDownloads/Publications/IRENA_RE_Latin_America_Policies_2015_Country_Mexico.pdf (6 September 2017).

IRENA. 2016. “Renewable Capacity Statistics 2016”. http://www.irena.org/DocumentDownloads/Publications/IRENA_RE_Capacity_Statistics_2016.pdf (19 July 2017).

Jacobsson, Staffan, and Volkmar Lauber. 2006. “The Politics and Policy of Energy System Transformation: Explaining the German Diffusion of Renewable Energy Technology”. *Energy Policy* 33 (4): 256–76.

Jeon, Sujin. 2011. “Overseas Oil Development of South Korea”. https://eneken.ieej.or.jp/data/3697.pdf (29 April 2018).

Jeong, Hyung S. 2010. “Prospect of Korea Climate Change”. *Rural and Environmental Engineering Journal* 109: 22–30.

Jones, Randall S., and Byungseo Yoo. 2011. "Korea's Green Growth Strategy: Mitigating Climate Change and Developing New Growth Engines". *OECD Economics Department Working Paper* 798.

Jordan, Andrew et al. (eds). 2018. *Governing Climate Change: Polycentricity in Action?*. Cambridge: Cambridge University Press.

Kaladharan, Megha. 2016. "Renewable Energy in India: An Analysis of the Regulatory Environment and Evolving Policy Trends". http://www.cprindia.org/system/tdf/working_papers/RE%20Regulatory%20and%20Policy%20Framework%20_26%20Aug.pdf (1 May 2018).

Kaneko, Shinji, and Masato Kawanishi (eds). 2016. *Climate Change Policies and Challenges in Indonesia*. Berlin: Springer.

Kemp, René, Johan Schot and Remco Hoogma. 1998. "Regime Shifts to Sustainability through Processes of Niche Formation: The Approach of Strategic Niche Management". *Technology Analysis & Strategic Management* 10 (2): 175–95.

Kennedy, Andrew B. 2010. "China's New Energy-Security Debate". *Survival* 52 (3): 137–58.

Keohane, Robert O., and David G. Victor. 2011. "The Regime Complex for Climate Change". *Perspectives on Politics* 9 (1): 7–23.

Kern, Florian. 2011. "Ideas, Institutions and Interests: Explaining Policy Divergence in Fostering 'System Innovations' Towards Sustainability". *Environment and Planning C* 29 (6): 1116–34.

Khurana, Mani, and Sudeshna G. Banerjee. 2015. "Beyond Crisis: The Financial Performance of India's Power Sector". https://openknowledge.worldbank.org/bitstream/handle/10986/20527/924900PUB0978100Box385375B00PUBLIC0.pdf (1 May 2018).

Kidd, Steve. 2013. "Brazil: Where Is It Going in Nuclear?". http://www.neimagazine.com/opinion/opinionbrazil-where-is-it-going-in-nuclear (25 April 2016).

Kim, Baek J. et al. 2002. "Summer Monsoon Rainfall Patterns over South Korea and Associated Circulation Features". *Theoretical and Applied Climatology* 72 (1–2): 65–74.

Kim, Sungjin. 2015. *The Adoption of the Paris Agreement and South Korea's Response to the New Global Climate Regime*. Seoul: EIA.

Kim, Sung Y., and John A. Mathews. 2016. "Korea's Greening Strategy: The Role of Smart Microgrids". *Asia-Pacific Journal* 24: n.p.

Klare, Michael T. 2008. *Rising Powers, Shrinking Planet: The New Geopolitics of Energy*. New York: Holt.

Knorr, Kaspar et al. 2016. "Wind and Solar PV Resource Aggregation Study for South Africa". https://www.csir.co.za/sites/default/files/Documents/Wind%20and%20Solar%20PV%20Resource%20Aggregation%20Study%20for%20South%20Africa_Final%20report.pdf (12 November 2016).

Kong, Bo. 2011. "Governing China's Energy in the Context of Global Governance". *Global Policy* 2 (s1): 51–65.

Korea Energy Agency. 2015. *Annual Report 2014*. Yongin City: Korea Energy Agency.

Kostka, Genia. 2014. "Barriers to the Implementation of Environmental Policies at the Local Level in China". *World Bank Policy Research Working Paper* 7016.

Kuzemko, Caroline et al. 2016. "Governing for Sustainable Energy System Change: Politics, Contexts and Contingency". *Energy Research & Social Science* 12: 96–105.

Lajous, Adrian. 2014. "Mexican Energy Reform". http://energypolicy.columbia.edu/sites/default/files/energy/CGEP_Adrian%20Lajous_Mexican%20Energy%20Reform_Final.pdf (21 August 2017).

Lewis, Joanna. 2011. "Energy and Climate Goals of China's 12th 5 Year Plan". https://www.c2es.org/site/assets/uploads/2011/03/energy-climate-goals-chinas-twelfth-five-year-plan.pdf (25 April 2018).

Liu, Qiang et al. 2015. "Peaking China's CO_2 Emissions: Trends and Mitigation Potential". http://www.wri.org/sites/default/files/uploads/WRI15_OCN_Peaking_Emissions_v4.pdf (25 April 2018).

Liu, Zhui et al. 2012. "Embodied Energy Use in China's Industrial Sectors". *Energy Policy* 49: 751–58.

Lontoh, Lucky, Kieron Clarke and Christohpher Beaton. 2014. "Indonesia Energy Subsidy Review: Biannual Survey of Energy Subsidy Policies". https://www.iisd.org/gsi/sites/default/files/ffs_indonesia_review_i1v1.pdf (23 June 2017).

Lontoh, Lucky, Christopher Beaton and Kieron Clarke. 2015. "Indonesia Energy Subsidy Review: Biannual Survey of Energy Subsidy Policies". https://www.iisd.org/gsi/sites/default/files/ffs_indonesia_reviewi2v1_english.pdf (23 June 2017).

Luomi, Mari. 2014. "Sustainable Energy in Brazil: Reversing Past Achievements or Realizing Future Potential". *OIES Paper* 34.

Ma, Damien. 2015. "Rebalancing China's Energy Strategy". http://www.paulsoninstitute.org/wp-content/uploads/2017/01/PPEE_Rebalancing-Chinas-Energy-Strategy_Ma_English_R.pdf (25 April 2018).

Macdonald, Juli A. 2008. "Perspectives on India's Global Energy Activities: Benefits for Foreign Policy, and National Security Implications". In *Foreign Addiction: Assessing India's Energy Security Strategies*, edited by Michael Kugelman, 18–26. Washington, DC: Wilson Center.

Macedo, Isaias C., Joaquim E. Seabra and João E. Silva. 2008. "Greenhouse Gases Emissions in the Production and Use of Ethanol from Sugarcane in Brazil: The 2005/2006 Averages and a Prediction for 2020". *Biomass and Bioenergy* 32 (7): 582–95.

Madan, Tanvi. 2006. "India". https://www.brookings.edu/wp-content/uploads/2016/06/2006india.pdf (7 June 2018).

Madureira, Nuno L. 2014. *Key Concepts in Energy*. Heidelberg: Springer.

Maggio, Gaetano, and Gaetano Cacciola. 2012. "When will Oil, Natural Gas, and Coal Peak?". *Fuel* 98: 111–23.

Mandela, Nelson. 1993. "South Africa's Future Foreign Policy". *Foreign Affairs* 72 (5): 86–97.

Mares, David R. 2006. "Natural Gas Pipelines in the Southern Cone". In *Natural Gas and Geopolitics From 1970 to 2040*, edited by David G Victor, Amy M. Jaffe and Mark H. Hayes, 169–201. Cambridge: Cambridge University Press.

Marquard, Andrew. 2006. "The Origins and Development of South African Energy Policy". PhD diss., University of Cape Town.

Mathews, John A. 2012. "Green Growth Strategies: Korean Initiatives". *Futures* 44 (8): 761–69.

Maupin, Agathe. 2015. "Energy and Regional Integration: The Grand Inga Project in the DR Congo". In *A New Scramble for Africa?: The Rush for Energy Resources in Sub-Saharan Africa*, edited by Sören Scholvin, 53–70. Farnham: Ashgate.

McDonald, David. 2009. "Electric Capitalism: Conceptualising Electricity and Capital Accumulation in (South) Africa". In *Electric Capitalism: Recolonising Africa on the Power Grid*, edited by David McDonald, 1–49. London: Earthscan.

McKinsey. 2014a. "Ten Ideas to Reshape Indonesia's Energy Sector". http://www.mckinsey.com/~/media/mckinsey%20offices/indonesia/pdfs/ten_ideas_to_reshape_indonesias_energy_sector.ashx (10 March 2017).

McKinsey. 2014b. "A Tale of Two Mexicos: Growth and Prosperity in a Two-Speed Economy". http://www.mckinsey.com/~/media/McKinsey/Global%20Themes/Americas/A%20tale%20of%20two%20Mexicos/MGI_Mexico_Full_report_March_2014.ashx (11 August 2017).

McKinsey. 2017. "North American Shale Outlook to 2025". https://www.mckinseyenergyinsights.com/services/market-intelligence/reports/north-american-shale-oil-outlook/ (5 May 2018).

McKune, Craig. 2010. "Pebble Bed Modular Reactor Demonstration Plant Is Funded But Not Constructed". *South African Journal of Science* 106 (5–6): n.p.

Meidan, Michal. 2016. "The Structure of China's Oil Industry: Past Trends and Future Prospects". *OIES Working Paper* 66.

Meidan, Michal, Amrita Sen and Robert Campbell. 2015. "China: The 'New Normal'". https://www.oxfordenergy.org/wpcms/wp-content/uploads/2015/02/China-the-new-normal.pdf (25 April 2018).

Metz, Bert et al. 2005. "IPCC Special Report on Carbon Dioxide Capture and Storage". http://www.ipcc.ch/pdf/special-reports/srccs/srccs_wholereport.pdf (23 November 2016).

Mexico Oil & Gas Review. 2014. "Expert Interviews: Operational Targets of Mexico Energy Policy". http://mexicooilandgasreview.com/2014/interviews/interview14_1.html (6 September 2017).

Ministério de Minas e Energia [of Brazil]. 2007. "Plano Nacional de Energia 2030". http://www.epe.gov.br/PNE/20080111_1.pdf (20 June 2017).

Ministério de Minas e Energia [of Brazil]. 2010. "Plano Decenal de Expansão de Energia 2019". http://www.epe.gov.br/PDEE/20101129_1.pdf (20 June 2017).

Ministério de Minas e Energia [of Brazil]. 2011. "Plano Nacional de Eficiência Energética". http://www.mme.gov.br/web/guest/publicacoes-e-indicadores/plano-nacional-de-eficiencia-energetica (20 June 2017).

Ministério de Minas e Energia [of Brazil]. 2014a. "Plano Decenal de Expansão de Energia 2023". http://www.epe.gov.br/PDEE/Relat%C3%B3rio%20Final%20do%20PDE%202023.pdf (9 July 2015).

Ministério de Minas e Energia [of Brazil]. 2014b. "Plano Nacional de Energia 2050: Cenário sócio-econômico e demanda de energia". http://www.mme.gov.br/documents/10584/10745/PNE_2050_workshop_eco_dem_vf.pdf (20 June 2017).

Ministério de Minas e Energia [of Brazil]. 2016. "Balanço Energético Nacional". https://ben.epe.gov.br/downloads/Relatorio_Final_BEN_2016.pdf (5 January 2017).

Ministério do Meio Ambiente [of Brazil]. 2010. "Política Nacional sobre Mudança do Clima". http://www.mma.gov.br/clima/politica-nacional-sobre-mudanca-do-clima (20 June 2017).

Ministry of Energy and Mineral Resources [of Indonesia]. 2016. "Rencana Usaha Penyediaan Tenaga Listrik, RUPTL 2016–2025". http://www.djk.esdm.go.id/pdf/RUPTL/RUPTL%20PLN%202016-2025.pdf (27 June 2017).

Ministry of National Development Planning [of Indonesia]. 2013. "Framework for Nationally Appropriate Mitigation Actions". http://ranradgrk.bappenas.go.id/rangrk/admincms/downloads/publications/Indonesias_framework_for_Nationally_Appropriate_Mitigation_Actions.pdf (22 June 2017).

Ministry of Power [of India]. 2016. "Report of the Auditor and Comptroller General of India on the Restructured Accelerated Power Development and Reforms Programme". http://admin.indiaenvironmentportal.org.in/files/file/Power_Performance_Audit.pdf (1 May 2018).

Ministry of Trade, Industry and Energy [of South Korea]. 2014. "Korea Energy Master Plan: Outlook and Policies to 2035". http://www.motie.go.kr/common/download.do?fid=bbs&bbs_cd_n=72&bbs_seq_n=209286&file_seq_n=2 (27 April 2018).

Mitchell, Catherine. 2008. *The Political Economy of Sustainable Energy*. Basingstoke: Palgrave Macmillan.

Mittra, Sonali. 2011. "Small Hydro: Too Small for a National Mission?". *ORF Issue Brief* 34.

Mochtar, Zainal A. 2015. "The Constitutionality of Subsidies and Legal Issues Related to Fossil-Fuel Subsidy Reform in Indonesia". https://www.iisd.org/gsi/sites/default/files/ffs_indonesia_briefing_constitution_eng.pdf (23 June 2017).

Mohan, Aniruddh. 2016. "The Future of Nuclear Energy in India". *ORF Occasional Paper* 98.

Mohr, Steve H., and Geoffrey M. Evans. 2009. "Forecasting Coal Production until 2100". *Fuel* 88: 2059–67.

Mohr, Steve H., and Geoffrey M. Evans. 2010. "Long Term Prediction of Unconventional Oil Production". *Energy Policy* 38 (1): 265–76.

Morris, Mike, and Lucy Martin. 2015. "Political Economy of Climate-Relevant Policies: The Case of Renewable Energy in South Africa". *IDS Evidence Report* 128.

Müller-Kraenner, Sascha. 2008. "China's and India's Emerging Energy Foreign Policy". *DIE Discussion Paper* 15/2008.

Muylaert, Maria S., Corbiniano Silva and Cristiano Pires. 2008. "Land Use Change Sector Contribution to the Carbon Historical Emissions and the Sustainability: Case Study of the Brazilian Legal Amazon". *Renewable and Sustainable Energy Reviews* 13 (3): 696–702.

National Institute for Transforming India. 2016. "Report of the Expert Group on 175 GW RE by 2022". http://niti.gov.in/writereaddata/files/writereaddata/files/document_publication/report-175-GW-RE.pdf (1 May 2018).

National People's Congress. 2005. *11th Five-Year-Plan for Economic and Social Development of the People's Republic of China 2006–2010*. Beijing: Compilation and Translation Bureau.

National People's Congress. 2010. *12th Five-Year-Plan for Economic and Social Development of the People's Republic of China 2011–2015*. Beijing: Compilation and Translation Bureau.

National People's Congress. 2015. *13th Five-Year-Plan for Economic and Social Development of the People's Republic of China 2016–2020*. Beijing: Compilation and Translation Bureau.

NEA and IAEA. 2016. "Uranium 2016: Resources, Production and Demand". http://www.iaea.org/inis/collection/NCLCollectionStore/_Public/48/009/48009747.pdf (7 June 2018).

NDRC. 2007. "China's National Climate Change Programme". http://en.ndrc.gov.cn/newsrelease/200706/P020070604561191006823.pdf (25 April 2018).

NDRC. 2017. "China's Policies and Actions for Addressing Climate Change". http://en.ndrc.gov.cn/newsrelease/201711/P020171108521968689324.pdf (25 April 2018).

Never, Babette. 2014. *Knowledge Systems and Change in Climate Governance: Comparing India and South Africa*. London: Routledge.

Never, Babette, and Joachim Betz. 2014. "Comparing the Climate Policy Performance of Emerging Economies". *World Development* 59: 1–15.

NITI Aayog. 2015a. "Report on India's Renewable Electricity Roadmap 2030: Toward Accelerated Renewable Electricity Deployment". http://niti.gov.in/writereaddata/files/document_publication/RE_Roadmap_ExecutiveSummary.pdf (1 May 2018).

NITI Aayog. 2015b. *India Energy Security Scenarios, 2047: Version 2.0*. New Delhi: NITI Aayog.

NITI Aayog. 2017. "Draft National Energy Policy". http://niti.gov.in/writereaddata/files/new_initiatives/NEP-ID_27.06.2017.pdf (1 May 2018).

Nkomo, Jabavu C. 2005. "Energy and Economic Development: Challenges for South Africa". *Journal of Energy in Southern Africa* 16 (3): 10–20.

Norbrook, Nicholas. 2012. "Buried Treasure". *Africa Report* 38 (2): 78–79.

O'Donnell, Jill K. 2013. "Nuclear Power in South Korea's Green Growth Strategy: Green Growth Quarterly Update III-2013". https://www.cfr.org/content/publications/attachments/ROKGreenGrowthQuarterlyReport_June2013.pdf (29 April 2018).

Oliveira, Patrícia. 2014. "Problemas no Setor Elétrico Aumentam Custos e Exigem Mais Investimento em Fontes Renováveis". http://www12.senado.leg.br/noticias/materias/2014/02/20/problemas-no-setor-eletrico-aumentam-custos-e-exigem-mais-investimento-em-fontes-renovaveis (25 April 2016).

Olivier, Jos G., Klara M. Schure and Jeroen A. Peters. 2017. "Trends in Global CO_2 and Total Greenhouse Gas Emissions: 2017 Report". http://www.pbl.nl/sites/default/files/cms/publicaties/pbl-2017-trends-in-global-co2-and-total-greenhouse-gas-emissons-2017-report_2674.pdf (29 April 2018).

Padayachee, Vishnu. 2010. "Re-introducing the Minerals–Energy Complex". *Transformation: Critical Perspectives on Southern Africa* 71: 1–2.

Paeng, Jung K. 2010. "National Plan for Green Growth". https://home.jeita.or.jp/greenit-pc/activity/asia/file/korea.pdf (29 April 2018).

Paik, Keun W. 2018. "South Korea's Energy Policy Change and the Implications for Its LNG Imports". *OIES Paper* 132.

Pande, Varad. 2011. "India at Cancun: The Emergence of a Confident Dealmaker". *Economic & Political Weekly* 56 (6): 14–15.

Parikh, Kirit S. 2012. "Approaches to the Development of Renewable and Clean Energy in Brazil, China, Egypt, India and South Africa: Lessons of Emerging Countries". *ICRIER Policy Series* 13.

Park, Jaehee. 2016. "Report on Wind Power Industry 2015, South Korea". https://www.rvo.nl/sites/default/files/2016/03/Rapport%20Windenergie%20Zuid-Korea.pdf (29 April 2018).

Park, Soo H. et al. 2016. "Can Renewable Energy Replace Nuclear Power in Korea?: An Economic Valuation Analysis". *Nuclear Engineering and Technology* 48 (2): 559–71.

Patti, Carlos. 2015. "Brazil: An Emergent Nuclear Power". In *Emergent Brazil: Key Perspectives on a New Global Player*, edited by Jeffrey D. Needell, 261–74. Gainesville: University Press of Florida.

Patzek, Tadeusz W., and Gregory D. Croft. 2010. "A Global Coal Production Forecast with Multi-Hubbert Cycle Analysis". *Energy* 35 (8): 3109–22.

PEMEX. 2013. "Plan de Acción Climática de Petróleos Mexicanos y sus Organismos Subsidiarios". http://biblioteca.semarnat.gob.mx/janium/Documentos/Ciga/Libros2011/CD001637.pdf (9 January 2018).

PEMEX. 2016. "Plan de Negócios 2017–2021". http://www.pemex.com/acerca/plan-de-negocios/Documents/plannegocios-pmx_2017-2021.pdf (9 January 2018).

People's Republic of China. 2015. "Intended Nationally Determined Contributions". http://www4.unfccc.int/Submissions/INDC/Published%20Documents/China/1/China's%20INDC%20-%20on%2030%20June%202015.pdf (29 April 2018).

Pereira Lindoso, Diego et al. 2011. "Climate Change and Vulnerability to Draught in the Semiarid: The Case of Smallholder Farmers in the Brazilian Northeast". In *Climate Change in Brazil: Economic, Social and Regulatory Aspects*, edited by Ronaldo Seroa da Motta et al., 235–54. Brasília: Institute for Applied Economic Research.

Petrobras. 2016. "Relatório Estratégico: Plano de Negócios e Gestão 2017–2021". http://www.petrobras.com.br/pt/quem-somos/estrategia/plano-de-negocios-e-gestao (23 March 2017).

Pinguelli Rosa, Luiz, Christiano Pires and Maria S. Muylaert. 2009. "Biofuel Contribution to Mitigate Fossil Fuel CO_2 Emissions: Comparing Sugar Cane Ethanol in Brazil with Corn Ethanol and Discussing Land Use for Food Production and Deforestation". *Journal of Renewable and Sustainable Energy* 1 (3): n.p.

Planning Commission [of India]. 2002a. "Tenth Five-Year-Plan 2002–2007 I: Dimensions and Strategies". http://planningcommission.nic.in/plans/planrel/fiveyr/10th/volume1/10th_vol1.pdf (1 May 2018).

Planning Commission [of India]. 2002b. "Tenth Five-Year-Plan 2002–2007 II: Sectoral Policies and Programmes". http://planningcommission.nic.in/plans/planrel/fiveyr/10th/volume2/10th_vol2.pdf (1 May 2018).

Planning Commission [of India]. 2002c. "Tenth Five-Year-Plan 2002–2007 III: State Plans: Trends, Concerns and Strategies". http://planningcommission.nic.in/plans/planrel/fiveyr/10th/volume3/10th_vol3.pdf (1 May 2018).

Planning Commission [of India]. 2008a. "Eleventh Five-Year-Plan 2007–2012 I: Inclusive Growth". http://planningcommission.nic.in/plans/planrel/fiveyr/11th/11_v1/11th_vol1.pdf (1 May 2018).

Planning Commission [of India]. 2008b. "Eleventh Five-Year-Plan 2007–2012 II: Sector Social". http://planningcommission.nic.in/plans/planrel/fiveyr/11th/11_v2/11th_vol2.pdf (1 May 2018).

Planning Commission [of India]. 2008c. "Eleventh Five-Year-Plan 2007–2012 III: Agriculture, Rural Development, Industry, Services, and Physical Infrastructure". http://planningcommission.nic.in/plans/planrel/fiveyr/11th/11_v3/11th_vol3.pdf (1 May 2018).

Planning Commission [of India]. 2011. "Mid Term Appraisal: Eleventh Five Year Plan 2007–2012". http://planningcommission.gov.in/plans/mta/11th_mta/chapterwise/Comp_mta11th.pdf (1 May 2018).

Planning Commission [of India]. 2013. "Twelfth Five-Year-Plan 2007–2012 II: Economic Sectors". http://planningcommission.gov.in/plans/planrel/12thplan/pdf/12fyp_vol2.pdf (1 May 2018).

Presidency [of South Africa]. 2005. "Accelerated and Shared Growth Initiative: South Africa (ASGISA)". http://www.presidency.gov.za/download/file/fid/1055 (9 January 2018).

President of the Republic [of Indonesia]. 2006a. "Kebijakan Energi Nasional". http://www.batan.go.id/prod_hukum/extern/Perpres5_2006.pdf (27 June 2017).

President of the Republic [of Indonesia]. 2006b. "Rencana Percepatan Pembangunan Pembangkit Tenaga Listrik". http://storage.jak-stik.ac.id/ProdukHukum/Sekneg/71.pdf (27 June 2017).

President of the Republic [of Indonesia]. 2014. "Kebijakan Energi Nasional". http://prokum.esdm.go.id/pp/2014/PP%20Nomor%2079%202014.pdf (27 June 2017).

Price, Lynn, Xuejun Wang and Jiang Yun. 2010. "The Challenge of Reducing Energy Consumption of the Top-1,000 Largest Industrial Enterprises in China". *Energy Policy* 38 (11): 6485–98.

Price, Lynn et al. 2011. "Assessment of China's Energy-Saving and Emission-Reduction Accomplishments and Opportunities During the 11th Five Year Plan". *Energy Policy* 39 (4): 2165–78.

Proskuryakova, Liliana. 2017. "Energy Technology Foresight in Emerging Economies". *Technological Forecasting and Social Change* 119: 205–10.

Puttaswamy, Nagalakshmi, and Mohammed S. Ali. 2015. "How Did China Become the Largest Solar PV Manufacturing Country?". *STEP Working Paper* 2/2015.

PwC. 2014. "Oil and Gas in Indonesia: Investment and Taxation Guide". http://www.pwc.com/id/en/publications/assets/oil_and_gas_guide_2014.pdf (17 March 2017).

PwC. 2015. "Let's Energise: Meeting India's Growing Fuel Demand". https://www.pwc.in/assets/pdfs/publications/2015/lets-energise-meeting-indias-growing-fuel-demand.pdf f (1 May 2018).

PwC. 2017. "The Long View: How Will the Global Economy Order Change by 2050?". https://www.pwc.com/gx/en/world-2050/assets/pwc-the-world-in-2050-full-report-feb-2017.pdf (24 January 2018).

Ramana, Mahesh V. 2010. "The Future of Nuclear Power". https://casi.sas.upenn.edu/iit/ramana (1 May 2018).

Ramos Martins, Fernando. 2008a. "Solar Energy Scenarios in Brazil, Part One: Resource Assessment". *Energy Policy* 36 (8): 2853–64.

Ramos Martins, Fernando. 2008b. "Solar Energy Scenarios in Brazil, Part Two: Photovoltaics Applications". *Energy Policy* 36 (8): 2865–77.

Ramos Martins, Fernando, and Enio Bueno Pereira. 2011. "Enhancing Information for Solar and Wind Energy Technology Deployment in Brazil". *Energy Policy* 39 (7): 4378–90.

Raven, Rob et al. 2015. "Niche Construction and Empowerment through Socio-Political Work: A Meta-Analysis of Six Low-Carbon Technology Cases". *Environmental Innovation and Societal Transition* 18: 164–80.

Reischer, Rosalind. 2012. "China and India Unite on Energy". https://thediplomat.com/2012/07/china-and-india-unite-on-energy (1 May 2018).

Rennkamp, Britta, and Fernanda Fortes Westin. 2013. "Feito no Brasil? Made in South Africa?: Boosting Technological Development through Local Content Policies in the Wind Energy Industry". https://open.uct.ac.za/bitstream/item/19518/Rennkamp_Feito_no_Brasil_Made_in_2013.pdf (4 May 2018).

Republic of Indonesia. 2007. "National Action Plan Addressing Climate Change". http://theredddesk.org/sites/default/files/indonesia_national_action_plan_addressing_climate_change.pdf (22 June 2017).

Republic of Indonesia. 2009. "Climate Change Sectoral Roadmap". http://theredddesk.org/sites/default/files/indonesia-cc-synthesis-roadmap-dec09.pdf (22 June 2017).

Republic of Indonesia. 2015. "Intended Nationally Determined Contribution". http://www4.unfccc.int/submissions/INDC/Published%20Documents/Indonesia/1/INDC_REPUBLIC%20OF%20INDONESIA.pdf (22 June 2017).

Republic of Korea. 2015. "Intended Nationally Determined Contribution". http://www4.unfccc.int/ndcregistry/PublishedDocuments/Republic%20of%20Korea%20First/INDC%20Submission%20by%20the%20Republic%20of%20Korea%20on%20June%2030.pdf (29 April 2018).

REN21. 2013. "Renewables 2013: Global Status Report". http://www.ren21.net/Portals/0/documents/Resources/GSR/2013/GSR2013_lowres.pdf (10 March 2017).

REN21. 2015. "Renewables 2015: Global Status Report". http://www.ren21.net/wp-content/uploads/2015/07/REN12-GSR2015_Onlinebook_low1.pdf (25 April 2018).

REN21. 2016. "Renewables 2016: Global Status Report". http://www.ren21.net/wp-content/uploads/2016/05/GSR_2016_Full_Report_lowres.pdf (29 April 2018).

Roberts, Simon. 2007. "Patterns of Industrial Performance in South Africa in the First Decade of Democracy: The Continued Influence of Minerals-Based Activities". *Transformation: Critical Perspectives on Southern Africa* 65: 4–35.

Rochon, Emily. 2008. "False Hope: Why Carbon Capture and Storage Won't Save the Climate". http://www.greenpeace.org/international/Global/international/planet-2/report/2008/5/false-hope.pdf (23 November 2016).

Rodrigues, Roberto. 2015. "Biofuels: Energy for the World". In *Emergent Brazil: Key Perspectives on a New Global Player*, edited by Jeffrey D. Needell, 183–98. Gainesville: University Press of Florida.

Rogelj, Joeri et al. 2016. "Paris Agreement Climate Proposals Need a Boost to Keep Warming Well Below 2°C". *Nature* 534: 631–39.

Röhrkasten, Sybille, and Kirsten Westphal. 2013. "IRENA and Germany's Foreign Renewable Energy Policy: Aiming at Multilevel Governance and an Internationalization of the Energiewende?". *SWP Working Paper* FG 8 2013/01.

Rojas Cardenas, José C. et al. 2017. "Energy Efficiency in the Mexican Iron and Steel Industry from an International Perspective". *Journal of Cleaner Production* 158: 335–48.

Rong, Fang, and Rai Victor. 2011. "Coal Liquefaction Policy in China: Explaining the Reversal since 2006". *Energy Policy* 39 (12): 8175–84.

Rotmans, Jan, René Kemp and Marjolein van Asselt. 2010. "More Evolution than Revolution: Transition Management in Public Policy". *Foresight* 3 (1): 15–31.

Roy, Pallavi, and Jesika Briones. 2016. "Market Information Report: Mexico". https://www.marsdd.com/wp-content/uploads/2015/07/Going-Global-Mexico_FINAL-LATEST.pdf (19 July 2017).

Sadorsky, Perry. 2009. "Renewable Energy Consumption and Income in Emerging Economies". *Energy Policy* 37 (10): 4021–28.

Samsam Bakhtiari, Ali M., and Farzad Shahbudaghlou. 2000. "Energy Consumption in the Islamic Republic of Iran". *OPEC Energy Review* 24 (3): 211–33.

Sarmiento Gutiérrez, Maria B. 2011. "From CDM to Nationally Appropriate Mitigation Actions: Financing Prospects for the Brazilian Sustainable Development". In *Climate Change in Brazil: Economic, Social and Regulatory Aspects*, edited by Ronaldo Seroa da Motta et al., 139–56. Brasília: Institute for Applied Economic Research.

Sartori, Giovanni. 1994. "Compare Why and How: Comparing, Miscomparing and the Comparative Method". In *Comparing Nations: Concepts, Strategies, Substance*, edited by Mattei Dogan and Ali Kazancigil, 14–34. Oxford: Blackwell.

Scholvin, Sören. 2014a. "South Africa's Energy Policy: Constrained by Nature and Path Dependency". *Journal of Southern African Studies* 40 (1): 185–202.

Scholvin, Sören. 2014b. *The Geopolitics of Regional Power: Geography, Economics and Politics in Southern Africa*. Farnham: Ashgate.

Scholvin, Sören. 2015. "Energy from Across the Border: Explaining South Africa's Regional Energy Policy". In *A New Scramble for Africa?: The Rush for Energy Resources in Sub-Saharan Africa*, edited by Sören Scholvin, 71–92. Farnham: Ashgate.

Scholvin, Sören et al. 2017a. "Gateway Cities in Global Production Networks: Exemplified by the Oil and Gas Sector". *Unicamp Texto para Discussão* 307.

Scholvin, Sören et al. 2017b. "Plugging into Global Production Networks: Density, Distance, Division and the Local Context of Brazil's Oil and Gas Sector". *Unicamp Texto para Discussão* 317.

Schroth, Stephen T., Jason A. Helfer and Courtney D. Jude. 2012. "India". In *Encyclopaedia of Global Warming & Climate Change*, edited by S. George Philander, 756–58. London: Sage.

Seligsohn, Deborah. 2011. "The Transformation of China's Energy System: Challenges and Opportunities". http://www.wri.org/sites/default/files/seligsohn_testimony_energy_and_commerce_2011-04-04.pdf (25 April 2018).

Seligsohn, Deborah. 2016. "How China's 13th Five Year Plan Addresses Energy and the Environment". https://www.uscc.gov/sites/default/files/Deborah%20Seligsohn_Written%20Testimony%20042716.pdf (7 June 2018).

Seligsohn, Deborah, and Angel Hsu. 2011. "How Does China's 12th Five Year Plan Address Energy and the Environment?". http://www.wri.org/blog/2011/03/how-does-china%E2%80%99s-12th-five-year-plan-address-energy-and-environment (25 April 2018).

SENER. 2013. "Prospectiva del Sector Eléctrico 2013–2027". http://sener.gob.mx/res/PE_y_DT/pub/2013/Prospectiva_del_Sector_Electrico_2013-2027.pdf (9 July 2015).

SENER. 2014. "Programa de Desarrollo del Sistema Eléctrico Nacional: PRODESEN 2015–2029". https://www.gob.mx/cms/uploads/attachment/file/54139/PRODESEN_FINAL_INTEGRADO_04_agosto_Indice_OK.pdf (8 September 2017).

SENER. 2015. "Estrategia Nacional de Energía 2014–2028". https://www.gob.mx/cms/uploads/attachment/file/214/ENE.pdf (6 September 2017).

SENER. 2016. "Prospectiva del Sector Eléctrico 2016–2030". https://www.gob.mx/cms/uploads/attachment/file/177626/Prospectiva_del_Sector_El_ctrico_2016-2030.pdf (26 August 2017).

Sennes, Ricardo U., and Thais Narciso. 2008. "Brazil as an International Energy Player". In *Brazil as an Economic Superpower?: Understanding Brazil's Changing Role in the Global Economy*, edited by Lael Brainard and Leonardo Martínez Diaz, 17–54. Washington, DC: Brookings.

Seo, Kyong H., and Hyun J. Lee. 2011. *A White Book of Changma*. Seoul: Korea Meteorological Administration.

Seroa da Motta, Ronaldo. 2011. "The National Policy on Climate Change: Regulatory and Governance Aspects". In *Climate Change in Brazil: Economic, Social and Regulatory Aspects*, edited by Ronaldo Seroa da Motta et al., 33–44. Brasília: Institute for Applied Economic Research.

Sewell, William H. 1996. "Three Temporalities: Toward an Eventful Sociology". In *The Historic Turn in the Human Sciences*, edited by Terence J. McDonald, 245–80. Ann Arbor: University of Michigan Press.

Shaofeng, Chen. 2011. "Has China's Foreign Energy Quest Enhanced Its Energy Security?". *China Quarterly* 207: 600–625.

Sharda, Jai, and Tim Buckley. 2016. "India's Questionable Ultra Mega Power Plans". http://ieefa.org/wp-content/uploads/2016/08/India%E2%80%99s-Questionable-Ultra-Mega-Power-Plans-Viability-Issues-Continue-to-Complicate-New-Coal-Fired-Projects-August-2016.pdf (1 May 2018).

Sheinbaum, Claudia, Belizza J. Ruíz and Leticia Ozawa. 2011. "Energy Consumption and Related CO_2 Emissions in Five Latin American Countries: Changes from 1990 to 2006 and Perspectives". *Energy* 36 (6): 3629–38.

Shift Project. n.d. "Countries with Highest Primary Energy Consumption per Capita". http://www.tsp-data-portal.org/TOP-20-Consumer-per-capita#tspQvChart (27 April 2018).

Shrimali, Gireesh, Sumala Tirumalachetty and David Nelson. 2012. "Falling Short: An Evaluation of the Indian Renewable Certificate Market". https://climatepolicy initiative.org/wp-content/uploads/2012/12/Falling-Short-An-Evaluation-of-the-Indian-Renewable-Certificate-Market.pdf (1 May 2018).

Singh, Daljit. 2016. "Competition and Choice in Electricity Distribution in India". http://www.cprindia.org/system/tdf/working_papers/Daljit%20Singh_Competition %20and%20Choice_24Aug.pdf (1 May 2018).

Smith, Adrian, and Rob Raven. 2012. "What Is a Protective Space?: Reconsidering Niches in Transitions to Sustainability". *Research Policy* 41 (6): 1025–36.

Smith, Adrian, Andy Stirling and Frans Berkhout. 2005. "The Governance of Sustainable Socio-Technical Transitions". *Research Policy* 34 (10): 1491–510.

Solís, Juan C., and Claudia Sheinbaum. 2013. "Energy Consumption and Greenhouse Gas Emission Trends in Mexican Road Transport". *Energy for Sustainable Development* 17 (3): 280–87.

Song, Ranping et al. 2015. "Assessing Implementation of China's Climate Policies in the 12th 5-Year Period". https://www.wri.org/sites/default/files/15_WP_China_Climate_Policies_final-v2_0.pdf (25 April 2018).

Sosa, Gustavo S. 2015. "Climate Change Policy and Energy Reform: An Assessment of Mexico's Foreign Policy". *Latin American Policy* 6 (2): 240–54.

Sovacool, Benjamin K. 2009. "Reassessing Energy Security and the Trans-ASEAN Natural Gas Pipeline Network in Southeast Asia". *Pacific Affairs* 82 (3): 467–86.

Stuart, Alexandra. 2015. "Indonesia's Energy Policy: Challenges and Options". In *Indonesia: A Regional Energy Leader in Transition*, edited by Natalie Bravo et al., 7–21. Washington, DC: NBR.

Styan, James B. 2015. *Blackout: The Eskom Crisis*. Johannesburg: Ball.

Sukarna, Djadjang. 2012. "Development of Geothermal Policies, Potential Resources and Implementation Target". Presentation at the Indonesia EBTKE Conference, Jakarta, 18 July.

Sun, Guodong. 2010. "Coal in China: Resources, Uses, and Advanced Technologies". https://www.c2es.org/site/assets/uploads/2010/03/coal-china-resources-uses-advanced-coal-technologies.pdf (25 April 2018).

Susandi, Armi. 2009. "Integration of Adaptive Planning across Economic Sector". Paper presented at the NWP Workshop on Integration of Approaches to Adaptation Planning, Bangkok, 12–14 October.

Swain, Ashwini K. 2016. “Political Economy of Distribution Reforms in Indian Electricity”. http://www.cprindia.org/system/tdf/working_papers/rap-india-mappingpower-maharashtra-2017-july.pdf (7 June 2018).

Tanter, Richard. 2015. “The Slovakian ‘Inspirasi’ for Indonesian Nuclear Power: The ‘Success’ of a Permanently Failing Organization”. *Asian Perspective* 39 (4): 667–94.

Tianjie, Ma. 2016. “China’s 5 Year Plan for Energy”. https://thediplomat.com/2016/08/chinas-5-year-plan-for-energy (25 April 2018).

Turner, Mark et al. 2003. *Decentralisation in Indonesia: Redesigning the State*. Canberra: Asia Pacific Press.

Unruh, Gregory C. 2000. “Understanding Carbon Lock-in”. *Energy Policy* 28 (12): 817–30.

Unruh, Gregory C. 2002. “Escaping Carbon Lock-in”. *Energy Policy* 30 (4): 317–25.

Urpelainen, Johannes, and Thijs van de Graaf. 2015. “The International Renewable Energy Agency: A Success Story in Institutional Innovation?”. *International Environmental Agreements* 15 (2): 159–77.

Van Aken, Tucker, and Orion A. Lewis. 2015. “The Political Economy of Noncompliance in China: The Case of Industrial Energy Policy”. *Journal of Contemporary China* 95: 798–822.

Van de Graaf, Thijs. 2013a. *The Politics and Institutions of Global Energy Governance*. London: Palgrave Macmillan.

Van de Graaf, Thijs. 2013b. “Fragmentation in Global Energy Governance: Explaining the Creation of IRENA”. *Global Environmental Politics* 13 (3): 14–33.

Van den Bergh, Jeroen et al. 2007. *Evolutionary Economics and Environmental Policy: Survival of the Greenest*. Cheltenham: Elgar.

Vermeer, Eduardo B. 2015. “The Global Expansion of Chinese Oil Companies, Political Demands, Profitability and Risks”. *China Information* 29 (1): 3–32.

Villicaña Ortiz, Eunice et al. 2015. “Solar Energy Potential in the Coastal Zone of the Gulf of Mexico”. *Renewable Energy* 81: 534–42.

Viola, Eduardo, and Matías Franchini. 2017. *Brazil and Climate Change: Beyond the Amazon*. London: Routledge.

Wang, Can, Yuan Yang and Junjie Zhang. 2015. “China’s Sectoral Strategies in Energy Conservation and Carbon Mitigation”. *Climate Policy* 15 (s1): 60–80.

Wang, Jianwei, and Jing Zou. 2014. “China Goes to Africa, a Strategic Move?”. *Journal of Contemporary China* 90: 1113–32.

Wang, Lianhe. 2015. “Sino-US Cooperation on Climate Change”. *China International Studies* 50 (1): 96–100.

Wang, Qiang et al. 2014. “Natural Gas from Shale Formation: The Evolution, Evidences and Challenges of Shale Gas Revolution in United States”. *Renewable and Sustainable Energy Reviews* 30: 1–28.

Wang, Yunqing, and Qigui Zhu. 2015. “Energy Price Shocks, Monetary Policy and China’s Economic Fluctuations”. *Asian-Pacific Economic Literature* 29 (1): 126–41.

Ward, Hugh. 2008. “Liberal Democracy and Sustainability”. *Environmental Politics* 17 (3): 386–409.

Westphal, Kirsten, and Sybille Röhrkasten. 2016. "The G20 and Its Role in Global Energy Governance". In *Sustainable Energy in the G20: Prospects for a Global Energy Transition*, edited by Sybille Röhrkasten, Sonja Thielges and Rainer Quitzow, 12–18. Potsdam: IAAS.

Wijayatunga, Priyantha, Deb Chattopadhyay and Prem N. Fernando. 2015. "Cross-Border Power Trading in South Asia: A Techno Economic Rationale". *Asian Development Bank South Asia Working Paper* 38/2015.

Williams, Lisa. 2014. "China's Climate Change Policies: Actors and Drivers". https://www.lowyinstitute.org/publications/chinas-climate-change-policies-actors-and-drivers (25 April 2018).

Wilson, Stephen, Yufeng Yang and Jane Kuang. 2015. "China's Electricity Sector: Powering Growth, Keeping the Lights on and Prices Down". In *China's Domestic Transformation in a Global Context*, edited by Ligang Song, 175–210. Canberra: Australian National University Press.

Winkler, Harald. 2007. "Long Term Mitigation Scenarios: Technical Report". Report prepared for Prepared for the Department of Environment Affairs and Tourism, in possession of the authors.

Winkler, Harald, and Andrew Marquard. 2009. "Changing Development Paths: From an Energy-Intensive to Low-Carbon Economy in South Africa". *Climate and Development* 1 (1): 47–65.

Winters, Matthew S., and Matthew Cawvey. 2015. "Governance Obstacles to Geothermal Energy Development in Indonesia". *Journal of Current Southeast Asian Affairs* 34 (1): 27–56.

Woertz, Eckart. 2013. "The Domestic Challenges in the Saudi Energy Market and their Regional and Geopolitical Implications". https://www.files.ethz.ch/isn/175229/05a485f202440778052158eb7ef9808b.pdf (13 December 2018).

World Bank. 2007. "World Development Report 2008: Agriculture for Development". http://siteresources.worldbank.org/INTWDR2008/Resources/WDR_00_book.pdf (22 April 2016).

World Bank. 2009. "Investing in a More Sustainable Indonesia". http://www-wds.worldbank.org/external/default/WDSContentServer/WDSP/IB/2009/11/19/000333037_20091119011618/Rendered/PDF/507620v20Revis1ox0info10CEA1english.pdf (15 February 2017).

World Bank. 2013. "Turn Down the Heat: Climate Extremes, Regional Impacts, and the Case for Resilience". http://www.worldbank.org/content/dam/Worldbank/document/Full_Report_Vol_2_Turn_Down_The_Heat_%20Climate_Extremes_Regional_Impacts_Case_for_Resilience_Print%20version_FINAL.pdf (4 June 2018).

World Bank. 2014. "Electric Power Transmission and Distribution Losses". http://data.worldbank.org/indicator/EG.ELC.LOSS.ZS (25 April 2016).

World Bank. 2016. "State and Trends of Carbon Pricing". http://documents.worldbank.org/curated/en/598811476464765822/pdf/109157-REVISED-PUBLIC-wb-report-2016-complete-161214-cc2015-screen.pdf (25 April 2018).

World Nuclear Association. 2017a. "Nuclear Power in Indonesia". http://www.world-nuclear.org/information-library/country-profiles/countries-g-n/indonesia.aspx (8 January 2018).

World Nuclear Association. 2017b. "Nuclear Power in South Korea". http://www.world-nuclear.org/information-library/country-profiles/countries-o-s/south-korea.aspx (27 April 2018).

World Nuclear Association. 2018. "Nuclear Power in India". http://www.world-nuclear.org/information-library/country-profiles/countries-g-n/india.aspx (23 May 2018).

World Resources Institute. 2014. "6 Graphs Explain the World's Top 10 Emitters". https://wri.org/blog/2014/11/6-graphs-explain-world%E2%80%99s-top-10-emitters (20 June 2017).

Wright, Jarred et al. 2016. "Comments on the Integrated Resource Plan 2016 Draft". Paper presented at the Integrated Resource Plan 2016 public hearing, Johannesburg, 7 December.

Wu, Fuzuo. 2013. "China's Pragmatic Tactics in International Climate Change Negotiations: Reserving Principles with Compromise". *Asia Survey* 53 (4): 778–800.

Wübbeke, Jost. 2013. "China's Climate Change Expert Community: Principles, Mechanisms and Influence". *Journal of Contemporary China* 82: 712–31.

Xie, Lei. 2011. "China's Environmental Activism in the Age of Globalization". *Asian Politics and Policy* 3 (2): 207–24.

Xuegong, Sun, Guo Liyan, Zeng Zheng et al. 2012. "Market Entry Barriers for FDI and Private Investors: Lessons from China's Electricity Market". *ERIA Discussion Paper* 17/2012.

Yang, Xiaoliang. 2016. "U.S.–China Clean Energy Collaboration: Lessons from the Advanced Coal Technology Consortium". https://www.wri.org/sites/default/files/US_China_Clean_Energy_Collaboration_lessons_from_the_Advanced_Coal_Technology_Consortium_1.pdf (25 April 2018).

Yelland, Chris. 2008. "The Impact of Electricity Theft and Non-Payment on the Generation Capacity Crisis in South Africa". http://www.ee.co.za/article/the-impact-of-electricity-theft-and-non-payment-on-the-generation-capacity-crisis-in-south-africa.html (22 November 2016).

Yoon, Jong H., and Kwang H. Sim. 2015. "Why Is South Korea's Renewable Energy Policy Failing?: A Qualitative Evaluation". *Energy Policy* 86: 369–79.

Yusuf, Arief A., and Herminia Francisco. 2009. *Climate Change Vulnerability Mapping for Southeast Asia.* Singapore: Economy and Environment Program for Southeast Asia.

Yusuf, Ariel A., Elizabeth L. Roos and Jonathan M. Horridge. 2018. "Indonesia's Moratorium on Palm Oil Expansion from Natural Forests: Economy-Wide Impacts and the Role of International Transfers". *Asian Development Review* 35 (2): 85–112.

Zhang, Ming, Shuang Dai and Yan Song. 2015. "Decomposition Analysis of Energy-Related CO_2 Emissions in South Africa". *Journal of Energy in Southern Africa* 26 (1): 67–73.

Zhidong, Li. 2014. "Peak Coal in China: Rethinking the Unimaginable". In *China's Energy Crossroads: Forging a New Energy and Environmental Balance*, edited by Philip Andrews-Speed et al., 29–46. Seattle: NBR.

Zhou, Nan, Mark D. Levine and Lynn Price. 2010. "Overview of Current Energy Efficiency Policies in China". *Energy Policy* 38 (11): 6439–52.
Zhou, Nan et al. 2011. "China's Energy and Carbon Emission Outlook to 2050". http://eaei.lbl.gov/sites/all/files/lbl-4472e-energy-2050april-2011.pdf (25 April 2018).
Ziramba, Emmanuel. 2008. "The Demand for Residential Electricity in South Africa". *Energy Policy* 36 (9): 3460–66.

Index

8th Basic Plan for Electricity Supply and Demand, 6, 97, 174
11th Five-Year Plan [of India], 7, 52–53, 64
12th Five-Year Plan [of India], 53, 64, 185
12th Five-Year Plan [of China], 29, 34, 42, 44, 184
13th Five-Year Plan [of China], 35, 175

Accelerated Power Development and Reform Programme (APDRP), 54
Angra Power Stations, 111, 117
APDRP. *See* Accelerated Power Development and Reform Programme
APG. *See* ASEAN Power Grid
Arahan Power Station, 78
ASEAN Power Grid (APG), 80, 172

Belo Monte Dam, 116
biofuel, 15, 31, 57, 70, 78, 87, 92, 103, 107, 110–11, 122, 124–25, 126n2, 132–33, 171, 176, 180, 192; biodiesel, 78, 101, 171; bioenergy, 66, 78, 115, 132, 171; biomass, 30, 34, 47, 50, 75, 79, 94, 108, 113, 117–18, 121, 130, 132, 148, 153, 160, 169–71, 177, 179, 189; ethanol, 31, 78, 107, 110, 115–6, 121, 125, 135, 171, 183
blackout, 52, 61, 77, 153, 167. *See also* load shedding

Cahora Bassa Dam, 155, 172
Campos Basin, 109
Cantarell Field, 130
carbon capture and storage (CCS), 14, 19, 23, 27, 42–43, 64, 85, 87, 102–103, 122, 143, 145, 149, 161, 183–84
carbon dioxide emissions, 1, 4, 23, 32, 44–45, 66, 69, 71, 79, 86–89, 104–105, 125–26, 133, 141–43, 145, 149, 151, 161–62, 164, 170, 186–87, 191. *See also* greenhouse gas emissions
carbon tax, 21, 142, 145, 163–64, 185, *188*
carbon trade/carbon trading, 44, 185
CCS. *See* carbon capture and storage
Central Asian Gas Pipeline, 9, 40
Central Bangko Power Station, 78
China Fusion Engineering Test Reactor, 43
China National Offshore Oil Corporation (CNOOC), 37
China National Petroleum Corporation (CNPC), 37

clean energy certificate, 135, 138, 179–80; renewable energy certificate (REC), 58, 179;
renewable portfolio standard (RPS), 95, 100–101
Climate Change Vulnerability Index, *10*, 11, 25n2
CNOOC. *See* China National Offshore Oil Corporation
CNPC. *See* China National Petroleum Corporation
Coal India, 49, 56, 58–59, 177, 179, 182
cogeneration, 134–35, 138, 141, 143
comparative research, 1, 18, 24–25, 189
Connect Central Asia Policy, 9, 63
critical juncture, 21–22, 129, 134, 146, 168, 170, 177, 190. *See also* path dependence
curtailment, 35, 141, 189–90

deforestation, 69, 87–89, 123–26, 186–87, *188*
distribution loss, 48, 111, 132, 140, 167, 171, 189
Draft National Energy Policy, 7, 53–54, 56, 67, 174

Emissions Trading Scheme [of South Korea], 101, 104, 184
Energy Coordination Committee, 57, 180
energy efficiency, 3, 14–15, 18, 21, 28–29, 33–35, 49, 53, 56–58, 87, 95, 101–102, 122–25, 129, 133–34, 141–42, 149, 163, 168, 170, 175, 184–85, 191; fuel efficiency, 33, 44, 101, 142
energy intensity, 3, 5, 13, 15–16, 22–25, 32–34, 38–39, 57–58, 65, 91, 95–96, 101, 104, 124, 133–34, 149–50, 160, 163, 167, 175–76, 178, 181–82, 184–85, 191–92
Energy Master Plan, 6, 95–96, 98, 174
Energy Reform [of Mexico], 15, 137–40, 142–43, 181, *188*
energy theft/electricity theft, 48, 112, 132, 157, 171, 178
energy transition, xi, 1, 18–23, 69, 81, 91, 101, 137, 140, 146, 153, 163, 168–69, 189–92
Eskom, 4, 8, 18, 145–46, 148–49, 151–55, 156–60, 162–64, 168–69, 172, 178, 180, 182, 185

Fast-Track Electrification Plan, 7, 75, 77
feed-in tariff, 35, 58, 95, 101–102, 117, 135, 138, 179–80
Fukushima incident, 96, 100

Garabi-Panambi Hydroelectric Complex, 114
Gasoduto Bolívia–Brasil (Gasbol), 8, 110, 119
geothermal energy/geothermal power, 7, 17, 23, 69–71, 75–79, 83, 85–86, 89n3, 130, 132, 135, 140, 169–70, 181, 184
global governance, 13–15
Great Pipeline of the South, 113
green growth, 16, 91, 94–95, 98, 101–102, 104–105, 141, 171, 176, 183, 191
greenhouse gas emissions, 13, 15–16, 43–45, 46n3, 46n8, 52, 88, 92, 101, 104–106, 123–25, 142, 149, 162–64, 184, 186–87. *See also* carbon dioxide emissions

IIRSA. *See* Iniciativa para a Integração da Infraestrutura Regional Sul-Americana
INDC. *See* Intended Nationally Determined Contribution
Inga Dam, 8, 155
Ingula Pumped Storage Power Station, 148
Iniciativa para a Integração da Infraestrutura Regional Sul-Americana (IIRSA), 8, 114
Integrated Resource Plan, 6, 149–50, 152–53, 160, 165n1, 173–74

Intended Nationally Determined Contribution (INDC), 24, 27, 45, 54, 66, 88, 91, 105, 107, 123, 129, 141–42, 163, 167, 176, 186–87, *188*, 192
International Renewable Energy Agency (IRENA), 14, 132
International Thermonuclear Experimental Reactor (ITER), 43
IRENA. *See* International Renewable Energy Agency
Itaipu Dam, 111, 116
ITER. *See* International Thermonuclear Experimental Reactor

Jirau Power Station. *See* Madeira River Hydroelectric Complex
Jupiter Field, 108

Kazakhstan–China Pipeline, 9, 40
KEPCO. *See* Korea Electrical Power Corporation
Koeberg Power Station, 148, 151, 162
KOGAS. *See* Korea Gas Corporation
Korea Electrical Power Corporation (KEPCO), 99–100, 178
Korea Gas Corporation (KOGAS), 93, 97–99, 178
Korea National Oil Corporation, 97–98
Kudankulam Power Plant, 8, 48
Kudu Field, 8, 155
Kusile Power Station, 151

Laguna Verde Power Station, 132
Lake Toba Pumped Storage Power Station, 75
liberalisation, 15, 18, 37, 47, 60, 80, 84, 91, 98–100, 117, 129, 135, 137–40, 143, 155, 167, 175, 178–79, 181, 187
liquefied natural gas (LNG), 40, 50, 56, 60, 70, 73, 83–84, 86, 93, 98–99, 115, 146, 170–71, 184, 189
load shedding, 5, 77, 154, 157
lock-in, 21; nuclear fusion, 27, 31, 43, 46n7, 182

Madeira River Hydroelectric Complex, 114
Majuba Power Station, 154
Medupi Power Station, 151, 158
Mercosul Energy Ring, 114
minerals–energy complex, 17, 159–60, 182

National Action Plan on Climate Change [of India], 57
National Climate Change Programme [of China], 33
National Climate Change Response White Paper [of South Africa], 161, 163–64, 184
National Energy Policy [of Indonesia], 69, 77–79
National Energy Strategy [of Mexico], 137, 140
National Strategy for Green Growth, 95
new normal, 27, 32, 42, 44, 175, 191
Nu River Project, 39

Oil and Natural Gas Corporation (ONGC), 9, 59, 63–64
Oil India, 59
ONGC. *See* Oil and Natural Gas Corporation

Paris Agreement. *See* UN Climate Conference in Paris
path dependence, xi, 1, 17–22, 167, 169, 189. *See also* critical juncture
PDE. *See* Plano Decenal de Expansão de Energia
Pebble Bed Modular Reactor (PBMR), 162, 183
PEMEX. *See* Petróleos Mexicanos
Pertamina, 71, 83–84
Perusahaan Listrik Negara (PLN), 77, 81–84
Petróleo Brasileiro SA (Petrobras), 107–10, 119–23, 178, 180–81
Petróleos Mexicanos (PEMEX), 136–37, 139–41, 179, 182
Piratininga Power Station, 110

Plano Decenal de Expansão de Energia (PDE), 6, 115–16
Plano Nacional de Energia (PNE), 115
PLN. *See* Perusahaan Listrik Negara
Política Nacional sobre Mudança do Clima (PNMC), 124–25
power-purchase agreement, 37, 117, 152–53, 178–80
privatisation. *See* liberalisation
ProAlcool, 110, 171
Programa de Incentivo às Fontes Alternativas de Energia Elétrica (PROINFA), 117–18

regional cooperation, 8–9, 17, 40, 62, 107, 113–14, 118–19, 148, 154–56, 167, 172, 177
REIPPPP. *See* Renewable Energy Independent Power Producer Procurement Programme
Rencana Usaha Penyediaan Tenaga Listrik (RUPTL), 77
Renewable Energy Independent Power Producer Procurement Programme (REIPPPP), 152–53
Rosatom, 8, 78, 117, 162
RUPTL. *See* Rencana Usaha Penyediaan Tenaga Listrik

Santo Antonio Power Station. *See* Madeira River Hydroelectric Complex
SAPP. *See* Southern African Power Pool
Sasol, 149, 159–61, 164, 169, 183
Secunda Plant, 149, 161, 164
Sinopec, 37, 119
Southern African Power Pool (SAPP), 155
subsidies, 3–5, 17–18, 34, 38–39, 42, 47, 49, 56, 58–60, 69, 78, 80–83, 85, 87, 95, 98-99, 101, 118, 133, 136–37, 139, 167, 171, 177–79, 181–82, 191

TAGP. *See* Trans-ASEAN Gas Pipeline Infrastructure Project
Target Management Scheme, 101, 104, 184
thorium, 31, 54, 63–64, 183
Three Gorges Dam, 31, 116
Top-1,000 Programme, 33–34
Trans-ASEAN Gas Pipeline Infrastructure Project (TAGP), 9, 79–80, 172
transmission loss. *See* distribution loss
Tupi Field, 108

UN Climate Conference in Cancún, 65
UN Climate Conference in Copenhagen, 13, 44, 65–66, 88, 105, 123, 163
UN Climate Conference in Paris, 13–14, 18, 34, 54, 66–67, 88, 105, 126, 176, 187
unconventional oil and gas/ unconventional hydrocarbons, 2–3, 7, 15, 30, 41, 50, 73, 129, 137, 139–40, 143, 164, 187; fracking/ hydraulic fracturing, 143, 143n5, 147, 164; shale gas, 2, 16, 30, 50, 73, 109, 131, 143, 146–47, 150, 158, 164, 168, 187
uranium, 31, 50, 54, 93–94, 111, 117, 132, 147–48, 159

waste-to-energy/waste-to-heat, 33, 78, 175

Xiluodu Dam, 31

About the Authors

Sören Scholvin is Research Fellow at the Institute of Economic and Cultural Geography at the University of Hanover, and Associate Researcher at the German Institute of Global and Area Studies. His research interests are the energy policy of the emerging economies, regional development in South America and sub-Saharan Africa, and global value chains in the Global South. His publications include the article "South Africa's Energy Policy: Constrained by Nature and Path Dependency" (*Journal of Southern African Studies*, vol. 40) and the edited volume *A New Scramble for Africa?: The Rush for Energy Resources in Sub-Saharan Africa.*

Joachim Betz is Associate Researcher at the German Institute of Global and Area Studies and Professor Emeritus at the University of Hamburg. His research interests are the energy policy of the emerging economies, economic and political developments in South Asia, globalisation, and development policy. He has published "Comparing the Climate Policy Performance of Emerging Economies" (*World Development*, vol. 59, with Babette Never), "India's Turn in Climate Policy: Assessing the Interplay of Domestic and International Policy Change" (*GIGA Working Paper*, no. 190) and "The Reform of China's Energy Policies" (*GIGA Working Paper*, no. 216).

www.ingramcontent.com/pod-product-compliance
Ingram Content Group UK Ltd.
Pitfield, Milton Keynes, MK11 3LW, UK
UKHW020213290926
15470UKWH00001B/152

9 781498 53436